*World of Wildlife*

*World of Wildlife:* 8

# ANIMALS OF AUSTRALASIA

**ORBIS · LONDON**

From the original text by Dr Félix Rodríguez de la Fuente
Scientific staff: P. de Andres, J. Castroviejo, M. Delibes, C. Morillo, C. G. Vallecillo
English language version by John Gilbert
Consultant editor: Dr Maurice Burton
Creative director: Brian Innes

© 1970 Salvat S. A. de Ediciones, Pamplona    © 1971–1974 Orbis Publishing Limited, London WC1
Printed in Great Britain by Jarrold & Sons Ltd, Norwich                ISBN 0 85613 497 X

# Contents

**AUSTRALASIA**

Chapter 95: The Australasian region: remarkable corner of the earth — 1
Chapter 96: The red heartland of Australia — 15
Chapter 97: The kangaroo: symbol of pride and prejudice — 39
Chapter 98: Mammals of the Australian wood and forests — 63
Chapter 99: Birds of the Australian woodland and savannah — 87
Chapter 100: The raucous, radiant world of parrots — 129
   *Order: Psittaciformes* — 153
Chapter 101: Rare species in peril: the flesh-eating marsupials — 155
Chapter 102: Watery oases of an arid continent — 177

**ISLANDS**

Chapter 103: The oceanic islands: a panorama of evolution — 199
Chapter 104: Madagascar, isle of wondrous plants and animals — 227
Chapter 105: Land of the lemur, indri and aye-aye — 245
Chapter 106: The evolutionary curiosities of the Galapagos Islands — 259
Chapter 107: Paradise of the Pacific isles — 275
Chapter 108: The outposts of the Atlantic — 289

# Illustration acknowledgments

D. Baglin/NHPA: 159, 187
D. Baglin/Zentrale Farbbild Agentur GmbH: 64
H. Beste/Ardea Photographics: 36, 40, 119, 147, 154, 181, 195, 205
H. Bielfeld/Bavaria Verlag: 86
L. Bissell/N. Palmer: 263, 265
W. Bonatti/Mondadori Press: 190, 191
S. & K. Breedon: 5, 6, 8, 16, 19, 27, 57, 62, 72, 73, 76, 82, 90, 91, 112, 158, 161, 162, 165, 166
J. R. Brownlie/Bruce Coleman: 19, 30, 31, 53, 67, 68, 92, 136, 139, 207, 213
A. Cash/C. E. Ostman: 198, 217
Miguel Ángel L. Castaños: 12, 64, 187
Ernesto Cerra: 21, 32, 33, 35, 42, 43, 74, 80, 85, 94, 95, 97, 99, 105, 109, 114, 123, 124, 125, 127, 131, 150, 151, 153, 164, 178, 182, 188, 210, 220, 232, 233, 239, 243, 249, 250, 265, 271, 297
G. Chapman/Ardea Photographics: 90, 135, 142, 147, 150, 203
B. J. Coates/Bruce Coleman: 113, 141
Bruce Coleman: 13, 75, 76, 77, 88, 115, 121, 128, 132, 137, 138, 143
F. Collett/Ardea Photographics: 33, 38, 176, 181
CSIRO: 55
M. D. England/Ardea Photographics: 215
F. Erize: 137, 169, 179, 261, 272
F. Erize/Bruce Coleman: 41, 45, 47, 49, 50, 79, 171
Antonio Escudero: 3, 4
J. Gerbec/Zentrale Farbbild Agentur GmbH: 10, 11
J. Hancock/Bruce Coleman: 175
D. Hanley/Photo Researchers: 99
J. Hannebicque/Zentrale Farbbild Agentur GmbH: 228, 229, 253, 255
M. P. Harris/Bruce Coleman: 266, 267
G. Hausle/Jacana: 277
G. Holton/Photo Researchers: 93
E. Hosking: 111, 295
R. Kinne/Photo Researchers: 279
C. de Klemm/Jacana: 277
José Lalanda: 6, 7, 11, 24, 25, 48, 51, 53, 54, 56, 69, 89, 100, 103, 104, 110, 112, 130, 131, 133, 138, 140, 145, 156, 157, 163, 167, 180, 182, 194, 237, 240, 242, 248, 249, 250, 257, 270, 272, 273, 280, 281, 282, 286, 287
G. Laycock/Bruce Coleman: 284

G. Leavens/Photo Researchers: 51
E. Lingren/Ardea Photographics: 95, 101, 193
R. D. Martin/Bruce Coleman: 244, 246, 247
I. McPhail/Bruce Coleman: 214
P. Montoya/Jacana: 285
E. Muench/C. E. Ostman: 209, 282, 283
W. T. Müller/Roebild: 59
R. K. Murton/Bruce Coleman: 235
Okapia: 34, 57, 70, 75, 79, 81, 96, 172, 183
Ousoff/Jacana: 105
F. Park/Zentrale Farbbild Agentur GmbH: 13, 14, 37, 116, 117
L. Pelligrini: 250, 264, 268, 270, 271
F. Petter/Jacana: 241
G. Pizzey/Bruce Coleman: 2, 8, 25, 27, 28, 45, 61, 71, 84, 89, 102, 121, 125, 126, 144, 157, 166, 185, 195
Popperfoto: 37
E. Puigdengolas: 108
C. Ray/Photo Researchers: 221
G. R. Roberts/C. E. Ostman: 65, 73, 201, 204, 209, 211, 225
A. Root/Bruce Coleman: 258
S. Saavedra/Safoto: 299
A. Schilling/Bruce Coleman: 248
K. Scholz/Zentrale Farbbild Agentur GmbH: 288
W. Schraml/Jacana: 85
V. Serventy/Bruce Coleman: 40, 73
J. Six: 109
P. Slater/Photo Researchers: 126
Marcelo Socías: 122, 124, 170, 186, 234, 262
Carmen Solinas: 200
M. F. Soper/Bruce Coleman: 206, 211, 215, 218, 219, 222, 293
Souricate/Jacana: 230
J. X. Sundance/Jacana: 286
T. C. Taylor/Bruce Coleman: 44, 52
W. Taylor/Ardea Photographics: 91, 101
R. Tercafs/Jacana: 230, 231
R. Thibout/Jacana: 251
K. W. Tink/Ardea Photographics: 10, 11
A. Visgae/Jacana: 236
J. Wallis/Bruce Coleman: 9, 18, 196
Zentrale Farbbild Agentur GmbH: 22, 226
Cover picture by courtesy of the
Australian Information Bureau

Separated from continental land masses for millions of years, the mammals, birds and reptiles of Australia, New Zealand and the Pacific islands have evolved in the most strange and wonderful ways.

Australia is the land of marsupials—mammals which carry their undeveloped young in an abdominal pouch until they can fend for themselves. Of these the most familiar are the high-jumping kangaroos and wallabies, yet only recently have scientists discovered the truth about the amazing birth adventures of the 'joey', shown here in a series of astonishing photographs. Then there is the popular koala or Australian teddy bear, which feeds its young on semi-liquid pap and as an adult eats only the leaves of gum trees, avoiding certain forms which have proved to be poisonous.

This eighth volume of *World of Wildlife* introduces the reader to a host of extraordinary animals. The dry Australian interior is the home of grotesque yet relatively harmless reptiles with frightening names such as frilled lizard, bearded dragon and thorny devil; and it is in the desert that the amazing mallee fowl builds its sand nest, regulating the temperature according to time of day and season to assist incubation. The watery oases are the haunts of the strange platypus, an egg-laying mammal with a beaver-like tail and duck-like beak. New Zealand too is a sanctuary of rare animals, especially birds such as the kiwi, takahe, kea and kakapo. These and many other species of the Australian region truly rank high among the world's natural marvels.

# CHAPTER 95

# The Australasian region: remarkable corner of the earth

In these pages we are concerned only with the wildlife of Australia which, in zoogeographical terms, is itself part of the Australian or Australasian region, including New Guinea, Tasmania, New Zealand and the Pacific islands. It is an absorbing and unexpectedly dramatic part of the world biologically and whereas everyone knows something about the koala, the kangaroo, the lyrebird and the emu, the broader pattern of events leading to the appearance and development of these and other species is not so well known except to the specialist.

All this is so trite that it is almost worthless to set it on paper and yet it must be said because it leads us to the striking fact that on the smallest continent in the whole world, and on some of its neighbouring islands there has been preserved a fauna which is unique. The reptiles, the birds and many other groups of animals living in Australasia are remarkable and outstanding in their structure, habits and appearance. But there is nothing to compare with the Australasian mammals.

What has happened, and is still happening, in Australasia is no different from what has been occurring elsewhere in the world – a clash between evolutionary groups in which only the fittest species have survived.

When the isthmus of Panama took shape, two or three million years ago, mammals from North America crossed to South America and came into conflict with many endemic species. In the case of predators the conflict was direct and often bloody, but in general it was simply a slow, unrelenting struggle for habitats and food, ending in the destruction of the older populations and the triumph of the new arrivals. This was the type of confrontation which brought about the gradual disappearance of the dinosaurs, the extermination of a number of giant-sized mammals

Australia is renowned for its birdlife and flocks of budgerigars, seen here drinking, are familiar sights in many regions.

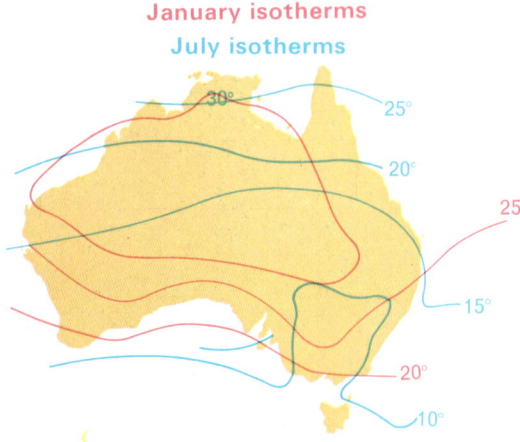

and the doom of several archaic animal forms; on a smaller and less spectacular scale it is this rivalry between old and new which is determining the pattern of animal life in modern Australia.

The colonisation of the continent by Europeans began with the landing by Captain James Cook on April 29, 1770, at Botany Bay—so named after the number of rare plants collected by the expedition. In the long run the arrival of the Europeans had its impact on the country's fauna. Two centuries later six species of mammals had been irreversibly decimated and twenty-eight others virtually made extinct. The earliest colonists, not surprisingly, were more interested in their own affairs than with the welfare of the local animal population. Those who did find time to study the country's wildlife were rewarded by the discovery of several wonderfully strange species. They included the kangaroo, the female of which carried her baby in an abdominal pouch; the duckbill or platypus, an aquatic animal with an otter-like body, a beaver's tail and a duck's beak, which laid eggs and incubated them; and the echidna or spiny anteater, looking like a long-snouted hedgehog and another egg-laying mammal. Nor could the nature lover be unaware of the remarkable beauty of the continent's birds which included magnificent parrots, parrakeets and cockatoos, and completely unfamiliar species such as black swans, lyrebirds, bowerbirds (so named after the elaborate bower-like structures built by the males for courting purposes) and the megapodes that fashioned incubating mounds out of sand and dead leaves.

Unfortunately the value of this remarkable zoological community was not fully appreciated at first. The white settlers started by killing the largest animals, the kangaroos, to make room for their imported sheep. Every conceivable method of slaughter was used, including poison. The 19th-century naturalist John Gilbert described how kangaroo-hunting developed into a popular sport. It was not unknown for the male kangaroos, instead of turning tail, boldly to face their attackers, supporting their backs, if possible, against a tree. They would then try to defend themselves against the hounds by lashing out with their powerful claws, kicking with their hind legs. An experienced pack would not come too close but would keep the kangaroos at bay, barking loudly until the hunters arrived. The latter would then proceed to beat out the animals' brains with sticks and clubs. Alternatively a kangaroo might make for nearby water, and a large male has been known to drag a pursuing dog under the surface, a manoeuvre that could only succeed if another dog did not plunge in to join the attack.

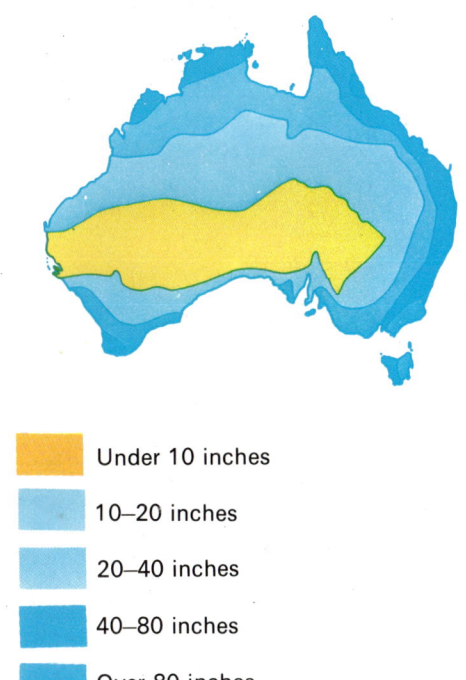

Australian rainfall chart (*above*); vegetation of Australian region (*below*).

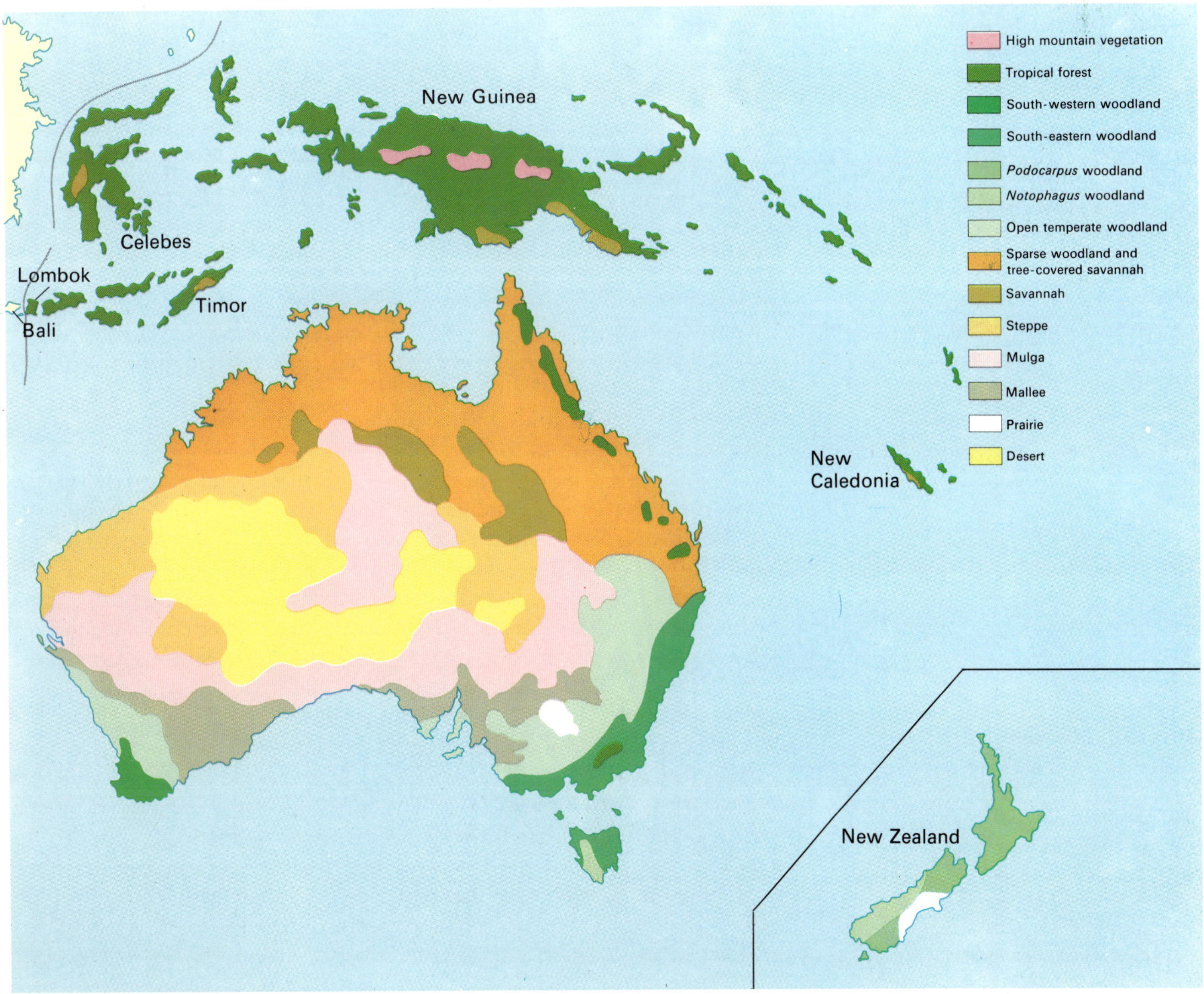

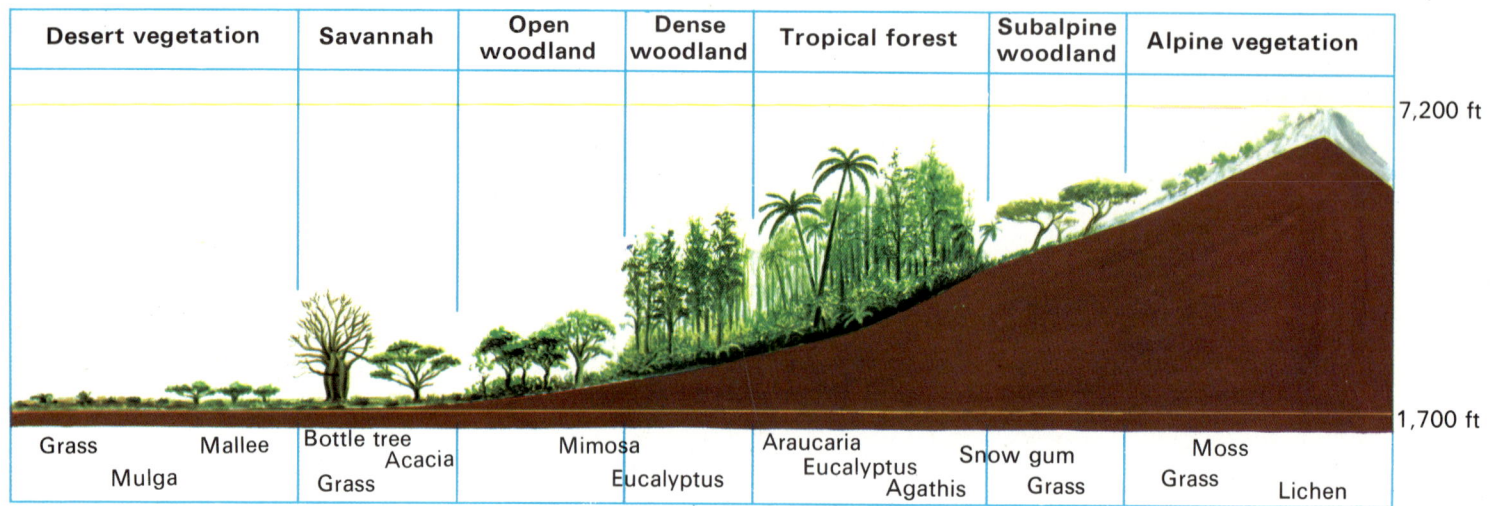

Diagram showing typical Australian vegetation, ranging from mountains to deserts of the interior. Humidity and altitude bring about diverse forms of plant growth, the tallest trees being found near the mountain chains where there is abundant rainfall. Increasing drought and flatness of terrain away from the coasts brings a hot, dry climate to the interior and lack of rain prevents trees and shrubs attaining any great size.

*Facing page:* The marsupials found their way to Australia before the placental mammals, populating almost every corner of the continent and becoming amazingly diversified. The largest marsupials are the kangaroos, whose range is nowadays more restricted than formerly, largely as a result of two centuries of intensive hunting. This picture shows a female wallaby protecting her 'joey'.

Man, however, was not the only menace to the kangaroo and to a host of smaller marsupials. They had to contend with a number of introduced species as well. Enormous flocks of sheep were let loose to graze on the plains; introduced rabbits, multiplying at a fantastic rate until their population was reckoned in tens of millions, overran immense areas of grassland, rendering it useless even for domestic livestock; and foxes, imported later to deal with the rabbit crisis, turned their attentions instead to the smaller members of the indigenous fauna. In addition dogs, cats and rats played havoc with small mammals and birds, spreading destruction through an animal community which had prospered in its seclusion from the outside world for millions of years.

## The earliest Australian mammals

Had the first arrivals from Europe made a thorough scientific investigation of the animals already inhabiting their island continent they would have been able to list no less than 104 species of placental mammals (Placentalia). All of them, apart from the dingo which was doubtless an introduced species, were comparatively small in size. They included forty-eight species of rats and mice belonging to the order Rodentia and fifty-one species of bats of the order Chiroptera.

Far more impressive (from the point of view of being larger and more diversified) were the Marsupialia, consisting of 124 species. These included jumping and climbing kangaroos; carnivores such as the Tasmanian wolf or thylacine, the Tasmanian devil, the native cat or dasyure, and the tiger cat; possums and phalangers; and, among smaller animals, marsupial mice, marsupial jerboas and marsupial moles.

The marsupials are a very ancient group of animals which today are represented outside Australia and New Guinea only by the Virginian opossum of North America and a variety of South American opossums. In addition to these, Australia is the home of the two remarkably interesting egg-laying mammals previously mentioned, the duckbill or platypus and the echidna, that date back over fifty million years ago, belonging to the order Monotremata.

These three branches of mammals were in competition with

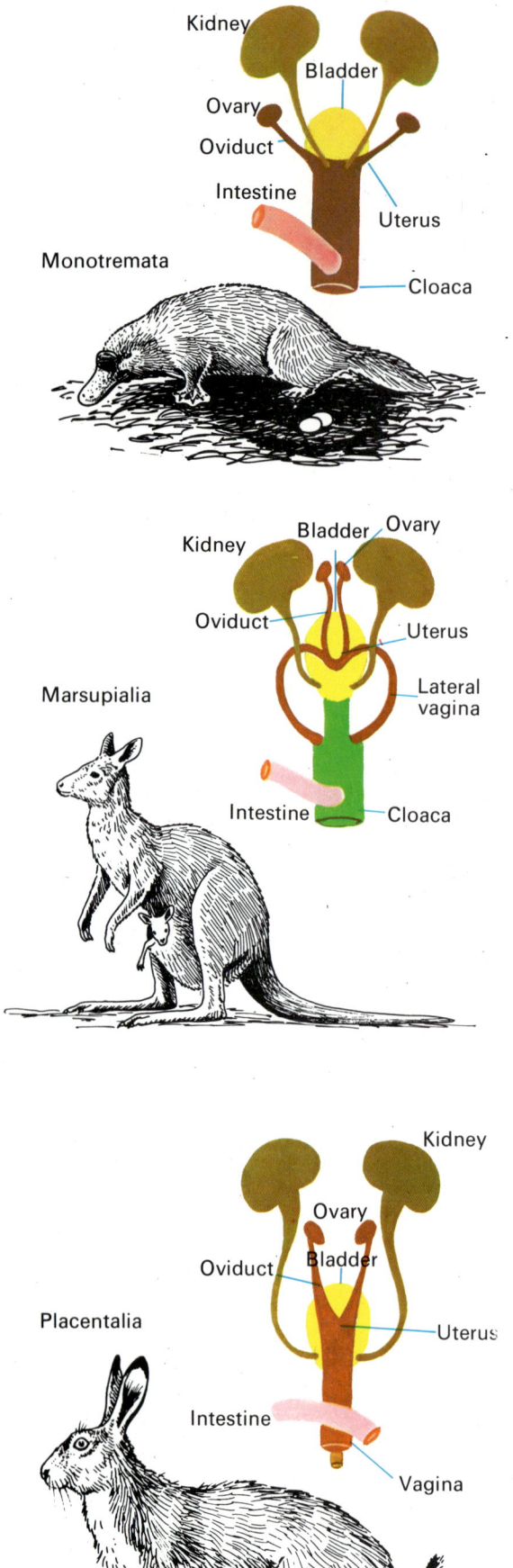

Monotremata

Kidney
Bladder
Ovary
Oviduct
Intestine
Uterus
Cloaca

Marsupialia

Kidney
Bladder
Ovary
Oviduct
Uterus
Lateral vagina
Intestine
Cloaca

Placentalia

Kidney
Ovary
Oviduct
Bladder
Uterus
Intestine
Vagina

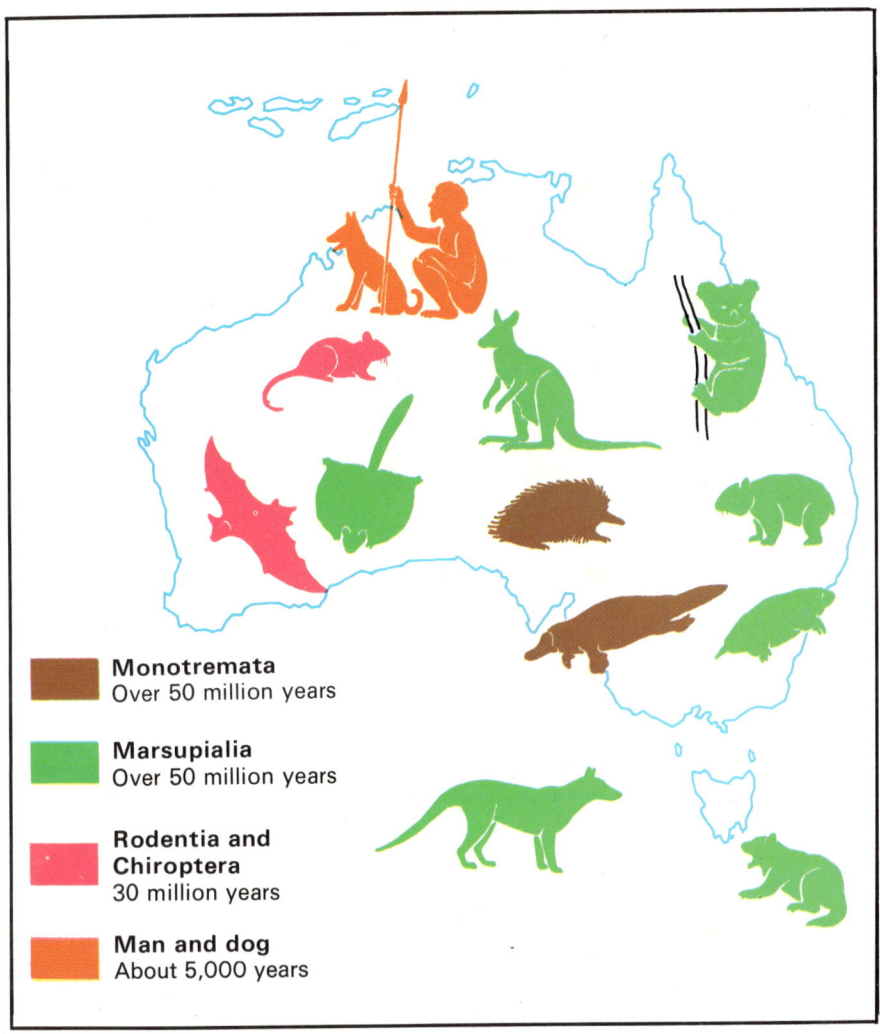

**Monotremata**
Over 50 million years

**Marsupialia**
Over 50 million years

**Rodentia and Chiroptera**
30 million years

**Man and dog**
About 5,000 years

The isolation of the Australasian region has made it difficult for true mammals to populate it. Before the white man introduced new species which destroyed the ecological balance of the region monotremes, marsupials and placental mammals all lived together. The monotremes, including the duck-billed platypus and five species of echidna, and the marsupials, comprising 124 species, are both descended from animals which arrived on the continent more than fifty million years ago. The earliest rodents are about thirty million years old and diversified into forty-eight species, while the bats, consisting of fifty-one species, reached Australia at about the same time, flying from island to island. The most recent arrivals – about 5,000 years ago – were primitive man and the dog.

one another in every part of the world at the beginning of the Tertiary epoch, taking over the domains formerly occupied by the giant reptiles. Each of them resolved the problems of reproduction in a different manner and only the most adaptable species managed to survive.

The Monotremata are in certain respects similar to the reptiles which they replaced, notably in the structure of their shoulder-girdle and in the possession of a cloaca (from which their name is derived), the outlet for the rectum and the genital and urinary tracts. They lack an efficient mechanism for the regulation of body temperature and, like reptiles, they too lay eggs although the mother incubates the eggs and suckles the young with secretions from her rudimentary mammary glands.

A cloaca, in anatomical terminology, is the single tube into which the bladder and intestine empty and through which the female receives the male's reproductive fluid and through which her babies or eggs, as the case may be, reach the exterior.

The Marsupialia have improved on the reptiles' reproductive system for they retain the embryos for a short time inside the body, the young being born while in the fœtal stage, at an early state of development, the subsequent development taking place inside the marsupial pouch, which is open to the outside and contains the mammary glands. Although the fœtus finds warmth and nourishment within the marsupium it is not entirely safe there,

*Facing page:* The essential anatomical difference between the monotremes, marsupials and placental mammals is the structure of the uterus, affecting the ways in which these animals reproduce. In the Monotremata it is rudimentary and non-functional; in the Marsupialia it is small and envelopes the embryo for the full marsupial gestation period; but in the Placentalia it is well developed. The genital and digestive tracts have a single outlet in the cloaca of the monotremes and marsupials but are independent in placental mammals. The marsupials also have two lateral vaginas which join to form a single vagina that enters the cloaca. The spotted cuscus *Phalanger maculatus* in the photograph is one of the many tree-dwelling marsupials.

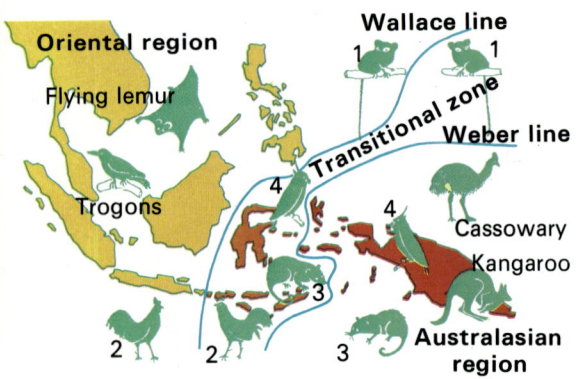

In 1863 Wallace traced an imaginary line to separate the Oriental from the Australasian region. A few years later Weber amended this division with a line farther east. In between the two lines are islands with a transitional flora and fauna. Although these islands sprang from the ocean some of their animal inhabitants have counterparts to the west and the east. Thus the tarsiers (1) and jungle fowl (2) found in the transitional zone are of Asiatic origin, whereas the cuscuses (3) and cockatoos (4) have come from Australia. The babirusa, a strange member of the pig family, illustrated on the opposite page, is only found in the transitional zone despite the fact that it has Asiatic ancestors.

*Preceding pages:* Among the more curious mammals of Australia are the Tasmanian devil (*above, left*), the echidna (*above, right*) and the duck-billed platypus (*below*). The largest snake of the Australian region is the rock python (*right*).

for if the mother is cornered by an enemy she may expel the baby from the pouch.

The Placentalia have evolved along another line. By means of the placenta the developing foetus is protected throughout the gestation period in the mother's belly and is fed by nutritious substances circulating in the bloodstream.

## Land links, past and present

Why was it that Australia became the last haven of the marsupials and monotremes? How did later arrivals to these shores such as the bats and rodents succeed in existing side by side with these indigenous species prior to the arrival of the white man? And where did the mammals of the Australasian region originate?

The islands making up the Australasian region cover an enormous area which is equivalent to the continent of Europe from the Straits of Gibraltar to the Urals. Although they all, to a greater or lesser degree, possess related forms of animal life, the various zoological communities are sharply differentiated from those of South-east Asia which we have just examined.

The Australasian region is separated from the Oriental region by the so-called Wallace-Weber lines (the former fixed by Alfred Russel Wallace, the latter by Max Weber), which follow the contours of the two deep sea troughs that separate the continental platforms on which Malaysia and Australasia rest. In between the two imaginary lines are islands with a transitional pattern of fauna—the Celebes, the Moluccas, Lombok, Flores and Timor.

To the north of the Wallace line the Philippines, and to the west Borneo, Java, Sumatra and the Malay peninsula retain typically Oriental features; and to the south-east of Australia the islands of New Zealand display so many individual characteristics and the geological reasons for the country's separation from Australia are so clear that most zoogeographers prefer to study it on its own.

From what we know of the submerged lands of South-east Asia and the Australian region it is reasonable to conclude that all colonisation of Australia originated in the Orient, by way of the straits and chains of islands separating the two zones. It has been proved that the levels of the inland seas of the two continental shelves fluctuated sufficiently at various times to permit the passage from west to east of many animal species. The only exceptions were in the very deep waters such as those that separate Bali from Lombok (along the Wallace line). Animal life may therefore be said to have flowed into the Australian region along a corridor corresponding to the arc formed by the islands of the Malay archipelago. The large island of New Guinea served as a kind of filter, as is evidenced by the fact that

Gleaming like a red jewel in the rays of the setting sun, Ayers Rock rises majestically above the surrounding plain. According to Aboriginal legend this was where the human race was fashioned and the clefts and ridges of the huge monolith testify to primitive habitation of the area. Today the surrounding region has been included in a National Park.

Babirusa
(*Babyrousa babyrussa*)

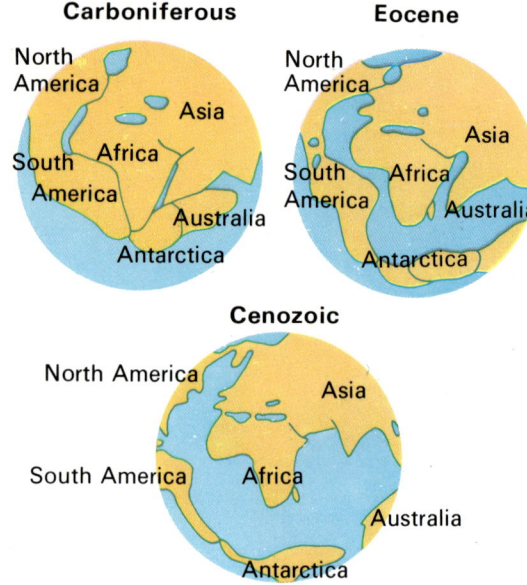

Although George Gaylord Simpson and many other scientists are convinced that the animals of the Australasian region originated in South-east Asia, crossing by land links, some of them currently submerged, there have been a number of other geological theories to explain the present-day distribution of flora and fauna. According to Alfred Wegener's theory of continental drift (*above*) all the continents were linked in the Carboniferous period and later began drifting apart. Thus the animals extant in the Carboniferous and Eocene periods would have been able to colonise Australia without encountering any ocean barrier. Another theory advanced by Edward Suess, based on the discovery of similar forms of reptiles, amphibians and plants in South America, Africa and Australia, suggests the existence of an enormous ancient continent—Gondwanaland—bounded to the north by the Tethys Sea. This would have permitted primitive animals to circulate freely prior to the subsequent breaking up of the continent into smaller land masses.

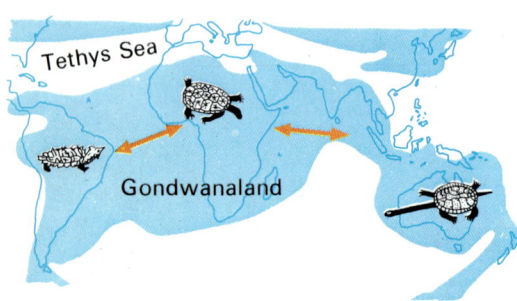

its vegetation consists principally of tropical rain forest, more reminiscent of South-east Asia than of Australia.

In any event geologists have now established that Australia has been separated from the Asiatic mainland at least since the beginning of the Tertiary, some seventy million years ago. In the view of George Gaylord Simpson, before the two continents drifted apart, even though the sole links were a narrow isthmus and a scattered chain of islands, marsupial mammals were distributed over both Eurasia and America. It was therefore pure chance that it was marsupials and not placental mammals that first reached the Australian continent.

There was no problem of dispersal for, except in the east where there was a mountain barrier, the land was flat, there were no earthquakes or volcanoes, and the climate (ranging from tropical and sub-tropical to temperate) did not fluctuate so suddenly and wildly as to force the species concerned to evolve a body temperature regulating mechanism. The existence of palms deep in the Australian deserts testifies to the fact that between the Pliocene and Pleistocene the climate must have been hot but wet, much wetter than today, for such trees could only have flourished where there was adequate rainfall. Furthermore, fossil remains of Diprotodontes—marsupials as large as rhinoceroses—and *Palorchestes* (kangaroos ten feet tall) indicate that these animals must have found luxuriant vegetation at their disposal. It was only much later that the climate changed dramatically and that low and irregular rainfall became the characteristic and predominant feature in so many areas. Today, apart from the eastern slopes of the mountains of south-eastern and eastern Australia as well as the south-western tip of the continent, and also the tropical north, for example, Cape York peninsula, which has up to 100 inches, the annual incidence of rainfall ranges from less than ten inches in the stony deserts of the interior to about twenty inches in the northern savannahs.

The marsupials therefore had little difficulty in adapting to life in different parts of the continent, some in the huge western shield or plateau, some in the central basin, others in the eastern highlands. They evolved in more or less the same manner as the placental mammals of other parts of the world, and with a number of striking resemblances. Thus the marsupial mice are difficult to distinguish at a glance from true rodents; the marsupial mole seems to be an almost exact copy of the golden mole from Africa; the marsupial rat-kangaroos look much like the kangaroo rats living in the North American deserts (but which are in fact rodents); and the Tasmanian wolf or thylacine, which feeds on live prey and carrion, puts one in mind of a true wolf. The phalanger glides, with the aid of its flying membrane, the patagium, in the same manner as the true flying squirrels of Eurasia and North America; and the lethargic marsupial known as the spotted cuscus moves about in the branches as indolently as a sloth. The only marsupials which have perhaps evolved in a different direction from placental mammals are the high-bounding kangaroos; the only similar animals are the much smaller kangaroo rats and jerboas.

After populating the continent the marsupials became

divided into two principal groups. Those in the extreme east and south-west and in the tropical north adapted to a predominantly warm, wet climate, while those in the central areas gradually accustomed themselves to the rigours of heat and drought.

The Monotremata were far fewer in number and did not offer serious competition. But later, after about thirty million years, the continent was invaded by rodents which proved just as adaptable as the marsupials, diversifying in remarkable fashion until there was a population made up of forty-eight species. The list of Australian mammals was completed at about the same period by bats which fluttered from island to island and adapted to their new surroundings without undergoing any significant transformations to distinguish them from the bats of the Oriental region.

During the expansion and radiation of these mammals there were predators present, such as the thylacine, Tasmanian devil and marsupial lion *Thylacoleo*, although the really damaging enemies arrived later still, approximately 5,000 years ago, in the shape of primitive man and the dog. The latter in particular was to have a dramatic impact on the wildlife of the continent. The dingo, returning to the wild, was partly responsible for the slow disappearance of the Tasmanian wolf on the mainland.

The thorny devil (*above*) is one of the strangest Australian reptiles. The dingo (*below*) was introduced by the aborigines.

# CHAPTER 96

# The red heartland of Australia

About one-third of Australia's land area is composed of arid land which becomes progressively drier the farther one travels into the interior. Moving across the great island from east to west one comes first to the eastern highlands, partially covered by evergreen rain-forest. This gives way to a pleasant landscape of bush, woodland clearings and savannahs dotted with occasional trees, the terrain becoming drier as one moves westwards. Next comes the mulga and mallee, a broad belt of flatland characterised by dwarf acacias and flanked in places by red sandstone plains. Farther west are steppe and desert; and it is in this great expanse of desert that we shall begin our survey of Australian flora and fauna, our focal point being the enormous natural dome known as Ayers Rock.

Ayers Rock is an immense stone monolith that rises more than 1,100 feet above the surrounding plain. It is two miles long and its base perimeter is about six miles. Impressive at any time of day, this solid slab of rock is unforgettable towards evening when the rays of the setting sun turn it an uncanny blood red. Little wonder that the rock has acquired legendary fame. The ancient Aborigines gave it the name Uluru—symbol of the earth itself.

According to Aboriginal legend, before any creature inhabited the earth the landscape was completely flat, without physical relief. It was giants from the underworld who fashioned the vast rock in the centre of the plain, and each cleft and hollow possesses special significance. Mysterious, unknown forces later created the human race which multiplied, divided into tribes and spread all over the continent. But they retained a distant memory of Uluru where they originated, returning there often to honour the sacred site, to pay homage to the subterranean

*Facing page:* The agamids of the Australian region are more bizarre in appearance than those found in other parts of the world and well deserve their popular names of devil and dragon.

Most of the Australian interior is characterised by an excessively hot, dry climate and for most of the year the landscape is punctuated only by small or stunted trees. Yet a brief period of rainfall will carpet the ground almost overnight with bright flowers, the seeds of which may have been lying dormant for years.

giants who fashioned them and to worship Wahambi, the snake spirit which judges mortals.

Today Ayers Rock is part of the Ayers Rock-Mount Olga National Park, established in 1958 – about 487 square miles of desert, containing a rich variety of plants and animals.

## Plants of the Australian desert

The Australian continent is affected by two systems of rainfall, one from the north, the other from the south. The former is the monsoon which brings relief to the northern parts of the country during the summer; the latter comes from the Southern Ocean, giving winter rain to the southern regions. As a result of one or other of these wet fronts rain-carrying clouds may drift across the continent at almost any time of year, bringing scattered showers that become increasingly rare towards the interior. But as happens in all the world's desert and semi-desert regions only a little rain is needed for the arid landscape to be magically transformed. Even the briefest shower may be sufficient to turn the expanse of red sand and stone into a carpet of multicoloured flowers. The seeds of these plants will perhaps have lain dormant for many years. The long-awaited moisture now results in the rapid growth of shoots, flowers and fruits in what is a short but glorious life cycle.

During the periods of drought, however, the colours in the desert are mainly reds, yellows and browns. One can travel for hundreds of miles seeing mainly mulga (*Acacia aneura*) and porcupine grass (*Triodia*), a monotonous pattern interrupted now and then by a thin line of dwarf eucalyptus (*Eucalyptus camaldulensis*) marking the course of a dried-up river bed, or the handsome silhouette of a she-oak (*Casuarina decaisneana*) which is usually found growing on its own or sometimes in small clumps. The heart of the desert is in fact composed almost wholly of mulga, on the fringes of which lie expanses of mallee, made up of stunted eucalyptus trees mixed with dwarf acacias and porcupine grass.

The traveller, particularly one who has had some first-hand experience of the African bush, may wonder why the Australian species of acacia do not bear long pointed thorns like their African counterparts. The reason is simply that the shoots and leaves of African acacias are continually nibbled by a variety of herbivores and have evolved such spines for protection. In Australia there are no equivalent species of large phytophages (herbivores) that browse leaves. On the other hand, porcupine grass, as its name suggests, is thorny.

Geographical distribution of the thorny devil.

## Devils and dragons: the strange reptiles of the desert

There has been much discussion and argument as to the origin of the Australian reptiles. Some zoological groups must certainly have come from Asia across the Indo-Australian archipelago; but there are others whose presence cannot be explained by this theory of continental links, now partially vanished. Whatever the explanation, Australia is the home of an extraordinarily rich community of desert reptiles. The richness of species is particularly noticeable in those reptiles which most people prefer to do without—the snakes.

Among the most curious of these reptiles are the snake-like lizards (Pygopodidae), of which there are ten species exclusive to Australia and an additional two found in New Guinea as well. These are lizards which in the long course of evolution have lost their forelimbs and whose hind legs are so reduced in size as to have lost their function of locomotion. Consequently the reptiles move in an undulating, snake-like manner.

Some of these strange species have a wide distribution in Australia and differ in certain details according to the region where they are found. Others are confined to particular zones, such as *Pletholax gracilis*, from Western Australia, and *Ophidiocephalus taeniatus* from South Australia, the latter very rare, only one captured specimen having been discovered back in 1896. Most of these lizards are insectivores but some feed on smaller lizard species.

The reptile family most closely related to these snake-like lizards is that which groups together the various geckos. These

The Queensland monitor, largest of twenty native Australian species, may grow up to eight feet long.

*Facing page:* The frilled lizard (*above*), measuring up to three feet, intimidates its enemies by opening its mouth and simultaneously spreading the erectile fold of skin on its neck. The thorny devil (*below*) is only about six inches long but covered from top to toe with sharp spines which are enough to deter most predators.

Gekkonidae are insect-eating lizards with nocturnal habits. Their tails are of different shapes and apparently store fat which can be utilised by the reptiles whenever food is in short supply.

Without any doubt, however, the most remarkable Australian reptiles are the Agamidae. It is true that agamids are also found in Asia, Africa and southern Europe, but here in Australia the appearance of some of them is sometimes so grotesque that one can readily understand how they have come to be called devils, dragons and similar names by the local inhabitants.

Best known of all is the thorny devil (*Moloch horridus*). Although only about six inches long this agamid is spectacular, its squat body being completely covered from head to tip of short tail with innumerable sharp spines. Larger spines crown the head above the snout and eyes like a pair of horns.

The thorny devil is an absolutely inoffensive reptile but it knows how to keep enemies at bay. Should there be any cause for alarm it will bury its head between its forelegs and puff out the swelling on its neck which is also protected by spines. It was once thought that this protuberance on the neck was a reservoir of fat but modern opinion is that it serves mainly to draw the attention of an enemy to a particularly well protected area of the body. Just as unusual as the external appearance of the thorny devil, however, are certain interesting biological features, such as its capacity for absorbing liquid through any section of its body. Dipping a part of the lizard's anatomy into water automatically sets off a series of swallowing movements, as if the animal were drinking. This astonishing phenomenon, which has been tested repeatedly under laboratory conditions, is brought about by the presence of a network of microscopically small canals which criss-cross the lizard's tough

Geographical distribution of the frilled lizard.

hide. These absorb water by capillary action and find an outlet in the reptile's mouth.

Another strange-looking Australian agamid is the frilled lizard (*Chlamydosaurus kingii*). This is a much larger reptile, about three feet in length. It too is harmless and if attacked will normally flee, running along on its hind legs only, tail held high, trying to find refuge in a tree. But if an enemy is too fast and corners it, the frilled lizard resorts to a highly effective method of intimidation. Standing steady on all fours it opens its mouth and lets out a whistling sound, at the same time spreading the huge frill encircling its neck. Opening the mouth and extending the frill are simultaneous actions made possible by the elongation of the hyoid bones which run from the tongue to the surface of the skin, arranged something in the manner of the struts of an umbrella.

A similar defensive apparatus comes into play in the case of the bearded lizard or bearded dragon (*Amphibolurus barbatus*). Here again a would-be predator is likely to be deterred by the same combination of gaping mouth and outspread collar.

The monitors of the genus *Varanus*, which we have already encountered in the Oriental region, have a broad geographical distribution throughout Australia. These large, muscular reptiles with their powerful teeth are known locally as goannas, probably a corruption of 'iguana', to which these agamids are related. The Aborigines find their flesh makes tasty eating – a useful source of protein in their diet. Twenty of the two dozen known species of monitors are Australian.

The largest member of the group is the Queensland monitor (*Varanus giganteus*), a desert inhabitant which may grow up to eight feet long. It seldom ventures far from its rocky haunts but in some parts of Western Australia it lives in trees.

All these monitors are both predators and scavengers, with a diet that includes insects, small mammals, snakes, lizards and birds.

Many goannas make use of termitaria in the breeding season. The female digs a tunnel in the termite mound and lays her eggs inside. The opening is immediately blocked up by the teeming army of insects hastening to repair the wall of their fortress. This method of reproduction is doubly advantageous for the monitor. In the first place there is a constant atmosphere inside the incubation chamber (both the temperature and the degree of humidity are subject to fewer variations than on the outside); and secondly the eggs are effectively protected from prowling or hovering predators.

Some authorities have advanced the theory that at the moment of hatching the baby reptiles that are born inside the termitarium break an exit hole in the surrounding walls and so manage to find their way outside. Others claim that it is the mother who scoops out the necessary aperture and helps her progeny to escape from their prison. Neither theory has been scientifically tested but the first seems the likelier of the two. It is rather difficult to accept the fact that the mother not only remembers precisely where she has deposited her eggs but also the exact moment when they are due to hatch.

Although the egg-laying activities of the female monitor are intriguing, it is also worth pausing to comment on the habits of the insects which play the unwilling hosts for the occasion. There are about 150 species of Australian termites, the most interesting of which are those belonging to the genus *Mastotermes*. These termites are of ancient stock for they were found in many parts of the world around the middle of the Tertiary epoch. But for some reason the majority of them later disappeared, the sole survivors being those that lived in Australia and which have modern counterparts there.

These termites differ in certain ways from those we have already examined in Africa. Thus instead of the termitaria taking the form of tall mounds or groups of columns (roughly circular or elliptical in cross-section) these constructions are flat. Although they may be more than 10 feet high and 8 feet wide the walls of the Australian termitaria are seldom more than 3–4 inches thick. Another curious feature is that the mounds are invariably built in a north-south direction. Of the several explanations advanced for this phenomenon the most likely seems to be that it has something to do with the regulation of interior temperature, reducing heat during the summer and making the most of the sunshine during the winter.

The largest of the Australian snakes is the amethystine or

If Africa and South America are renowned respectively for their large mammals and birds, Australia has a claim to be called the land of reptiles. Among many species are geckos, agamids and snake-like lizards.

North Queensland rock python (*Liasis amethystinus*) which sometimes measures more than 25 feet long. The smallest is the pygmy python, whose length barely exceeds 2 feet. There are nine or ten species of python, all of which are mainly nocturnal by habit and good climbers. Some of them feed principally on rabbits.

In other parts of the world the Colubridae are the most numerous and widely distributed of all snake families but in Australia they are sparsely represented. The few rare species, although venomous, are not highly dangerous because their teeth are situated in the lower portion of the mouth.

The Elapidae, all of which are poisonous snakes, are much commoner and in fact some 60 per cent of Australia's ground snakes belong to this family. Numerically speaking, Australia contains more venomous snakes than any of the other continents; and like the cobras, coral snakes and similar formidable species, a few of these Australian elapids are very dangerous to humans. However, they prey chiefly on small mammals, only occasionally attacking larger animals, and unless actually cornered these snakes prefer to flee rather than strike back at an enemy. This is true of snakes as a whole.

## The enigma of the desert birds

Every ornithologist knows that certain seasons are more suitable than others for watching birds and that some of the most spectacular flights of all (such as the flights of wild geese, the dances of cranes or the pre-migration assemblies of storks) can only be witnessed in particular places and for very brief periods. In studying bird behaviour so many factors have to be taken into account, not the least of which are weather and climate. And if there is one area in the world where the ornithologist has to be especially vigilant it is in the Australian desert where conditions may change dramatically in a matter of a few hours.

In the red heartland of Australia a heavy shower is a rare and wonderful occurrence. It is certainly not an annual event and may only happen at ten-year intervals. When for a short time the desert is miraculously transformed into a flower-strewn plain, birds that have not been sighted for years flock into the area to breed, resulting in an abnormal population increase. Among such enigmatic species, for example, are the chats of the genus *Ephthianura,* which at one time had not made an appearance for more than twenty years. In exceptional circumstances, when food happens to be plentiful for two consecutive years, Australian naturalists note with delight that certain species thought to have been extinct are still flourishing.

Such occurrences are unfortunately few and far between. Dearth and drought inevitably return. The plants wither as quickly as they have appeared, the ponds and puddles dry up and the desert soon reverts to normal—an endless expanse of rock and sand with a scattering of stunted trees. Many of the birds that visited the area for breeding disappear once more, arousing new fears that this time they will never come back.

Rainfall plays such a vital role in determining the pattern

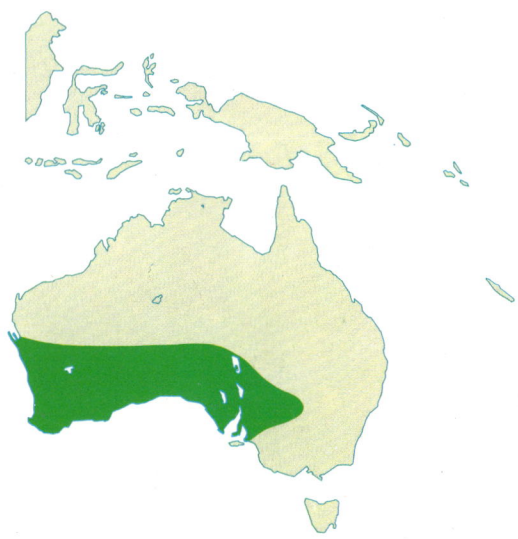

Geographical distribution of the mallee fowl.

*Facing page:* Australian termites construct mounds several feet high and wide but only a few inches thick. They always face from north to south, perhaps to reduce the sun's heat in summer and intensify it in winter.

---

**MALLEE FOWL**
(*Leipoa ocellata*)

Class: Aves
Order: Galliformes
Family: Megapodiidae
Diet: insects and other small animals; seeds, fruit
Number of eggs: up to 35
Incubation: 50–80 days, usually about 60 days

Colour greyish with numerous brown spots on back; line of dark feathers on front of neck and breast. Strong legs and claws. Chicks born with some feathers, enabling them to fly within twenty-four hours.

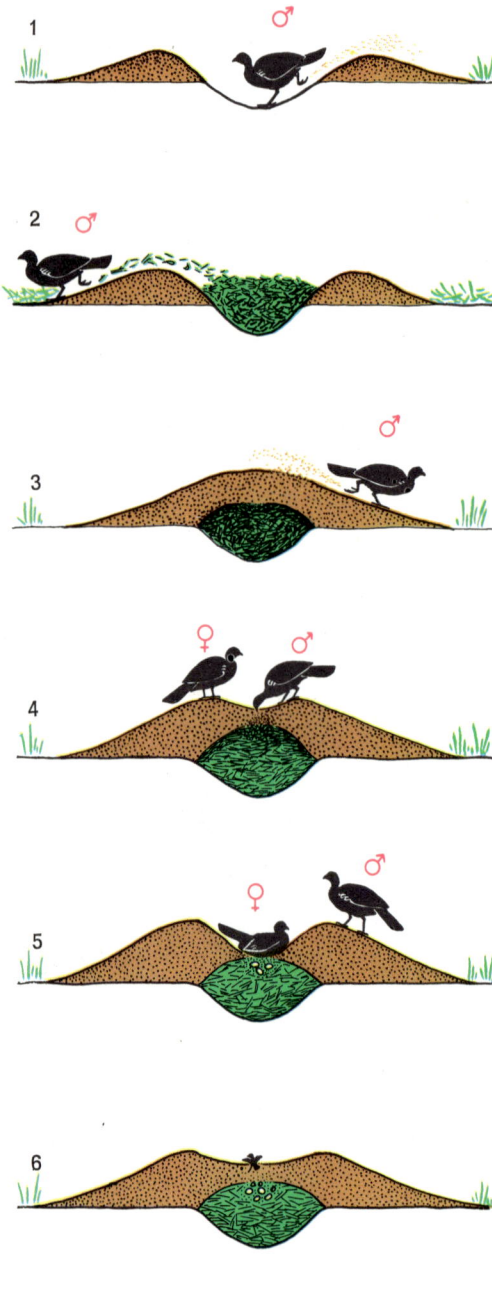

The male mallee fowl is occupied for about eleven months of the year in building a nest and incubating the eggs. He begins by scooping out a hole (1) which he then fills with leaves and twigs (2). After this pile of vegetation has been moistened by rain he shovels on a layer of sand (3). After testing the temperature of the rotting vegetation inside the mound with his beak (4) he lets the female lay her eggs (5). The eggs are incubated by the heat released by the rotting vegetation and when the chick hatches it has to make its own way to the outside (6).

of birdlife in the Australian desert that it is a matter of luck rather than planning to be in exactly the right place at the right moment. But there are certain species which are not directly influenced by rainfall incidence and which happen to be among the most interesting birds on the continent. They include mallee fowl, emus, falcons and wedge-tailed eagles.

## The mystery of the sand mounds

The earliest white settlers who ventured into the Australian interior expected to find the terrain completely flat and featureless and so for the most part it was. But they were surprised to come across, from time to time, mounds of sand, standing up to 5 feet high and measuring 15–20 feet in diameter. At first they assumed that these hillocks were primitive funeral monuments; but when they questioned the Aborigines they were astonished to be told that the mounds had not been built by human hands and that far from concealing the remains of a dead chieftain or witch-doctor they were actually birds' nests.

It is hardly surprising that this information should have been greeted with incredulity. Europeans had their own fixed and conventional ideas about what a bird's nest should look like and these huge sand castles offended all logic. So they decided to treat the Aborigines' statements as false, or at least as highly doubtful. It was left to the naturalist John Gilbert in 1840 to put the matter to the test in the only sensible way, by breaking open the mounds. He soon realised that the Aborigines had been right for inside them he found piles of large eggs, subsequently identified as those of the mallee fowl (*Leipoa ocellata*), a scrub savannah bird measuring more than 2 feet long and weighing 4–5 lb.

The mallee fowl is a species of megapode, strange birds which do not incubate their eggs by their own body heat, or protect their nestlings. The family name is derived from a Greek word meaning 'large foot' and these ground birds are indeed characterised by their exceptionally large, strong feet and claws. Some lay their eggs in a rock cleft, allowing them to incubate in the natural heat of the sun. Others, living on volcanic islands, take advantage of the heat emanating, directly or indirectly, from lava. But the mallee fowl, an inhabitant of mallee eucalyptus scrub regions (hence its name) follows the custom of forest megapodes, excavating a hole and filling it with vegetation which, as it rots, provides the necessary heat for embryo development, so acting as an incubator. There is, however, an essential difference, occasioned by the climatic variations between forest and desert. A forest megapode does not have to be much concerned about temperature fluctuations inside its mound of leaves; but the mallee fowl has to busy itself for as much as eleven months of the year with its incubation chamber to make sure that the eggs are not roasted in summer or frozen in winter as a result of the extreme climatic variations of its habitat. Although the incubation period of a single egg is two to two and a half months, the female lays up to 35 eggs at intervals and this means constant supervision on the part of the male.

## Problems of mound incubation

The method whereby the Megapodiidae deposit their eggs in places where the sun, volcanic sand or rotting vegetation provide the requisite heat for the developing embryos dispenses with the chores of normal incubation. But the system only really saves time and energy in regions, such as the tropical forest, where temperature variations are minimal. In temperate zones, where there are extreme fluctuations during the day and from season to season, it involves an expenditure of a great deal more energy than would be occasioned by the standard form of incubation from body heat.

The almost incessant labours of the mallee fowl have been meticulously studied and described by the Australian biologist Dr. H. J. Frith and it is on his reports that the following information is largely based.

It is the male who begins the preparations for the nest about four months before the female is due to lay her first egg. He starts his work by scooping a circular hole in the sand with his powerful feet. The cavity is 6–7 feet across and 3–4 feet deep.

When the female mallee fowl is ready to lay an egg the male inserts his bill into the mound of sand to test the temperature inside. If suitable he allows her to proceed; if not she has to delay the act. This happens repeatedly for she is capable of laying up to 35 eggs at intervals of four days.

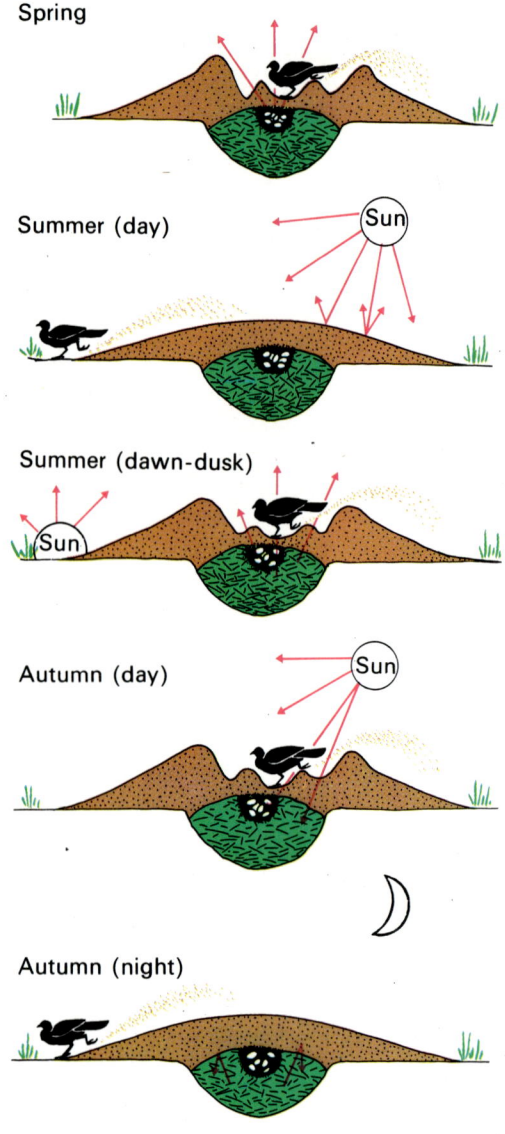

*Above and facing page:* Although building the nest and surrounding mound is arduous work, the male mallee fowl's problems really begin after the eggs are laid. In order to keep the interior temperature constant at about 33°C he has to keep opening entrances in the sand to allow more of the sun's heat to penetrate (as in spring, on summer evenings and by day during the autumn) or to close them (as on a summer's day to keep out the sun's rays or on an autumn night to retain the heat which has accumulated in the daytime). In this way the eggs can be incubated, whatever the season, either by means of the sun's natural heat or by the heat produced by the heap of rotting vegetation, in which they are embedded, or by a combination of both.

When this is done he stacks up all available scraps of vegetation within a radius of about 50 yards and crams it into the hole. How fast he works depends on atmospheric conditions. If it is dry he proceeds at a steady, unhurried rate; but if the weather is wet he puts on a spurt. In any event once he has got the pile into position he is unable to continue his work until it rains, for the heap of leaves, twigs and sticks will not start rotting until thoroughly wet. Once this happens the mallee fowl covers the entire nest with a thick layer of soil or sand.

Having done this, however, his problems are far from over. Now comes the most difficult task of all—maintaining the temperature inside the nest at around 33°C, with a maximum permissible fluctuation of one or two degrees. There are two possible heat sources available—the rotting vegetation and the warmth of the sun. Neither of these will be uniform so that the mallee fowl is compelled to be active continuously, according to the time of day and the season, for the seven months or so that may elapse between the laying of the first egg and the hatching of the last.

When the female indicates that she is ready to lay an egg, the male approaches the mound and burrows with his beak into the sand to test the temperature of the rotting vegetation below. How this natural thermometer works is not known but it is possible that he uses the tongue for the delicate operation. If satisfied with his findings he permits his mate to deposit the egg; if not he prevents her. He repeats this testing activity each time she prepares to lay an egg. This may occur up to 35 times at four- to eight-day intervals from mid-September to mid-February. This seems remarkable until we remember that many animals are highly sensitive to particular temperatures, although the way they detect them has seldom been investigated.

## Not too hot, not too cold

During the seven months when the nest is occupied, the male returns to the mound repeatedly to test the temperature and to carry out any necessary repairs. In the spring the sun is not too strong and all the heat for the eggs must derive from the rotting vegetation in which they are embedded. The bird assists the process by scooping holes in the surrounding layer of sand so as to hold the temperature steady. In summer, however, the sun is the principal source of heat and the problem is how to keep the eggs sufficiently cool. This the mallee fowl does by heaping up additional sand around the nest. In autumn the decomposition of the vegetable material slows down because there is little moisture and the sun's rays are still powerful. Consequently the male removes the protective layer of sand during the day to allow the heat to get to the eggs and scuffs it all back in the evening so that there is a minimal loss of heat which has accumulated in the daytime.

In order to make a more detailed study of the reactions of the mallee fowl to the temperature variations inside the incubation chamber Frith decided to heat a nest artificially at different times of the year. When he raised the interior temperature during the

The young mallee fowl is born with some of its wing and tail feathers so that once it has broken out of its mound it can take refuge in nearby thickets and flutter short distances to avoid predators. At no stage in its life will it see its parents and must run all the risks of a solitary existence until mature enough to find a partner and build a nest.

spring (the season when no natural heat entered from outside) he noted that the bird simply increased the number of openings in the sand to allow fresh air to circulate and cool the nest. But when he provided extra heat in the summer (at a time when incubation would normally have been effected only by the sun) the bird, unable to trace the source of the additional warmth, reacted by increasing the thickness of the sand covering. Similarly, in autumn, sensing that the temperature was about right, the mallee fowl omitted to open up the mound in the usual manner around midday to expose the eggs to the feeble rays of the sun.

When the temperature was deliberately increased or decreased, resulting in unseasonable conditions inside the nest, the bird was clearly perplexed. In such situations the female would sometimes join him in adjusting the sand level but on some occasions they would be working against each other, unsure about how to proceed. In the end, however, it was the male who guided the constructional work back to the right path.

In another experiment Frith himself built a nest to determine exactly how the internal temperature was affected by the various seasonal activities of the bird. To his great surprise he noticed that all his efforts were being attentively watched by the male mallee fowl who would visit the site several times a day to see how things were getting on. The bird would even scoop piles of sand to and fro as if to rectify the scientist's miscalculations!

The operations involved in supervising the incubation of the mallee fowl's eggs are so complex that the first attempts by younger, inexperienced males to build and maintain a nest may be utter failures. Several years will elapse before they manage to protect the eggs to the point of hatching. But once the male has mastered the skills of mound building the percentage of births is very high.

The only serious threat to the mallee fowl is the fox, introduced from abroad. Frith examined 1,094 eggs from 70 nests. He found 15 that were broken, 130 that had failed to hatch and 407 which had been eaten by foxes. Despite these losses the fact remained that 70 pairs of birds had given birth to 949 chicks, necessitating a high mortality rate among the offspring for the population to remain stable.

## The lonely fledgling

Because the adult mallee fowls spend some eleven months preparing and maintaining their nest they have little time to spare for looking after the young. So the latter have to fend for themselves. In fact they do not ever see their parents. When an egg hatches on its soft layer of vegetation, buried in a deep layer of sand, the nestling's first task is to find a means of breaking through to the surface. Otherwise its cradle soon becomes its tomb. This entails on its part many hours of energetic activity. Once out, the baby bird wanders away from the mound and seeks refuge in the nearest bushes where it leads a solitary existence until mature enough to find a partner. Its principal means of survival as it grows is its ability to run; this it can do as soon as it is born. Furthermore, within twenty-four hours or so it is capable of flying. Since it is not born with a covering of down on its body, like other baby birds, but is already provided with some wing and tail feathers, it is able to flutter short distances and thus to escape certain predators.

## Continental coincidences

Each of the world's zoogeographical regions has its own characteristic fauna, with a number of species that are not found anywhere outside its boundaries. Nevertheless these regions often contain natural habitats that resemble one another, such as the prairie of North America, the pampas of South America, the steppes of central Asia and the savannahs of Africa. Because of these similarities there are sometimes interesting parallels between the zoological communities of different continents. These groups of animals fulfil the same types of ecological role in their respective regions. Thus although the only ungulates in North America are the bison and the pronghorn, their place in the food chain is the same as that of the numerous ungulates of the African savannahs. Sometimes there are strong resemblances between the species of different regions, one of the most obvious and striking examples being among the ratites or flightless birds, represented by the African ostrich, the South American rhea and the Australian emu.

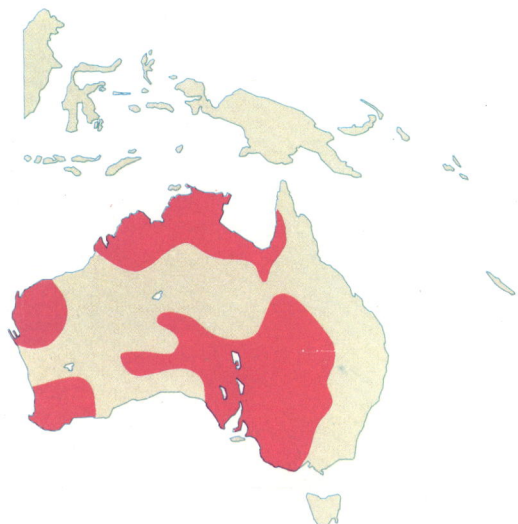

Geographical distribution of the emu.

---

**EMU**
(*Dromaius novaehollandiae*)

Class: Aves
Order: Casuariiformes
Family: Dromaiidae
Height: up to 72 inches (183 cm)
Weight: up to 120 lb (55 kg)
Diet: vegetation, insects
Number of eggs: up to 15
Incubation: 52–63 days

Second largest bird in world, after ostrich, but whereas latter has two digits on either foot, the emu has three. The feathers appear double, like those of cassowary, the aftershaft being almost the same length as the main feather. The young have cream plumage with longitudinal brown stripes.

---

*Following pages:* The emu, largest of Australian birds, is flightless, like the related cassowary, leading a nomadic life in which its main concern is to find sufficient vegetation and water.

The ostrich, the emu and the rhea provide a striking example of convergent evolution. Although superficially alike they are not related; but all three have adapted to similar conditions on three different continents. Scientists do not know whether they stem from an ancient common ancestor.

These three birds with their long legs, serpentine neck and small head look so similar to one another that the layman would perhaps take it for granted that they belonged to the same family. In fact the resemblance is only skin-deep, so much so that ornithologists classify them in three separate orders. The apparent similarity is thus probably a case of convergent evolution whereby the three species, although inhabiting different environments, have been subjected to parallel pressures and have adapted to conditions in the same way. All three have become so tall that they have lost their powers of flight and acquired long legs which have turned them into rapid runners.

It should be recognised, however, that science has not said its last word on this fascinating subject. We do not know for certain whether these adaptations appeared independently on the three continents concerned or whether the birds in question once had a common flightless ancestor.

## The emu: Australia's largest bird

The emu (*Dromaius novaehollandiae*) is second only in size to the African ostrich. As such it is Australia's largest bird, ranging widely over almost the entire continent. A related species, the cassowary (*Casuarius casuarius*) replaces the emu in the northeastern coastal forests and is also found in New Guinea and the adjoining islands. The emu has today retreated from the larger centres of population but may still be encountered on the outskirts of many towns.

Prior to the arrival of the European colonists emus were also prevalent on Tasmania, and King and Kangaroo Islands, but they were soon wiped out, being hunted for their flesh, their fat and their large eggs. The birds were hunted no less enthusiastically by the Australian Aborigines. Pictures of emus have been found etched on cave walls and carved into tree bark. As has been the case with other primitive people elsewhere, the animal in its role of hunting trophy has become a part of folklore. Among some tribes the emu appears as a totem and its gait and various distinctive movements are faithfully reproduced in local dances.

Colonisation did not have so disastrous an impact on the emu

Emu
(*Dromaius novaehollandiae*)

as it did on other species. Despite the fact that in 1865 it proved necessary to take measures to prevent a serious decline in numbers, the recovery was such that in this century the emu has become a serious nuisance in certain areas.

These swift-running birds are nomadic. Except in the course of the breeding season when they form pairs, they move about in small bands in search of food and water. When they find an area which satisfies their requirements they come together to form enormous flocks. Thanks to their long, powerful legs they can travel at around 30 miles per hour over fair distances. Nor are rivers a problem for the birds are accomplished swimmers.

The emu has a varied diet, consisting basically of leaves, grass, fruit and (especially in winter) large quantities of insects. The stomach of one bird was found to contain approximately 3,000 caterpillars, each measuring a couple of inches long. (It is interesting to calculate that if placed end to end this string of insects would have stretched to more than 150 feet!) As is the case with the ostrich, curiosity sometimes leads the emu to swallow a strange miscellany of inedible objects too, such as coins, keys, bottle tops and the like.

Cultivated crops are also much favoured, to the intense annoyance of farmers. Each year, shortly before the harvest, birds that have until then wandered at random through the deserts of the Australian interior converge on the fields of golden wheat, battering down fences and ruining acres of cereal either by eating it or trampling it underfoot. They also take over drinking troughs especially provided for livestock. In many places farmers have

An emu egg and chick.

Emus are typical birds of the Australian desert and are still found in large numbers in spite of having been hunted both by the Aborigines and white settlers.

Emus normally wander about in search of food in small groups; but when they come across plentiful supplies – as in wheat fields in harvest time – they gather in flocks and cause considerable damage.

persuaded the authorities to take protective measures. In Western Australia, where it is officially regarded as a scourge, 285,000 emus were killed between 1945 and 1970, the Ministry of Agriculture offering four shillings a beak; and in a two-year campaign in Queensland 121,768 birds were destroyed, together with over a million eggs.

A mass slaughter of emus was planned in 1932 when prominent local residents in the wheat growing area of Campion, Western Australia, brought pressure on the government to help them by sending soldiers to cope with the large flocks of emus that were destroying their crops. As a result a detachment from the Royal Australian Artillery engaged the 'enemy' with machine guns near Campion. It was estimated that there were approximately 20,000 birds in the region and the hope was that the army would achieve an easy victory. Unfortunately the emus refused to be sitting targets and decided to scatter in all directions, so that the soldiers were obliged to change tactics, resorting to traps and ambushes. This guerilla warfare proved as disappointing as the original grand offensive and after a couple of months of desultory skirmishing the army returned to barracks.

## Breeding and rearing

When the time comes for breeding the small flocks break up and the birds disperse to form pairs. The female emits a muffled cry which is echoed by the male. Then both birds stand side by side, heads drooping. Eventually the female sinks to the ground and the male mates with her.

The female is responsible for building the nest, scooping a hole in the ground and filling it with leaves and grass. She lays up to 15 dark green eggs, each of which measures $5\frac{1}{2}$ x 3 inches and weighs $1\frac{1}{2}$ lb. The eggs are laid in February or March (autumn in Australia) and incubation lasts 52–63 days. It is the father who devotes himself to this task, so assiduously in fact that he may spend all these weeks without eating or drinking, losing up to 20 lb in weight. In zoos. incidentally, this does not normally happen. During the incubation period the male will not object to someone touching or moving the eggs but when the young are born he will defend them staunchly. The emu chicks weigh a little over a pound and are cream with dark longitudinal stripes. At two months they are already more than a foot high and they remain with their parents until they are one and a half years old. At two years they are sexually mature.

## Desert birds of prey

There are a number of raptors in Australia's desert regions. Some of them are local representatives of species that have a global distribution, such as the black kite, which drifts with the wind above the scrub-covered plains, or the marsh harrier, which nests as close as possible to a river or swamp, this being the focal point of its hunting activities. Other birds of prey have their counterparts in the arid zones of Asia, Africa and America; and a few are sufficiently versatile to be at home both in the desert and other biomes, a prime example being the wedge-tailed eagle (*Aquila audax*), notable for being the most savagely persecuted eagle in the world.

This bird of prey is found on all types of terrain with the exception of forest and other areas of dense vegetation, for it operates best in open country. During the breeding season a couple of wedge-tailed eagles will fiercely defend the territory surrounding their eyrie but in general the birds are tolerant of others of the species. To some extent these raptors hunt much as vultures do in other parts of the world. Climbing to altitudes that render them invisible to the naked eye, they survey an enormous area. When one of them sights the body of an animal below it dives down, immediately followed by its companions. Several birds may then gather round to partake of the feast.

Although food consists to some degree of carrion these eagles also hunt many live mammals and birds. Ever since the introduction into Australia of the rabbit this animal has been a favourite prey, replacing small wallabies. Although lambs are also occasionally killed damage in this respect is to some extent compensated by their hunting of crop-destroying rabbits, which is a point in the eagle's favour.

Geographical distribution of the Australian kestrel

**AUSTRALIAN KESTREL**
(*Falco cenchroides*)

Class: Aves
Order: Falconiformes
Family: Falconidae
Wing-length: male $9\frac{1}{2}$–10 inches (24–25·5 cm)
female 10–$10\frac{3}{4}$ inches (25·5–27·5 cm)
Diet: basically insects, also small birds and mammals
Number of eggs: 4–6
Incubation: 26 days

Head and neck of male grey with some black streaks; back pale chestnut with black specks; tail black with white tip; breast and belly whitish. Female has no grey on head and more black marks on back; tail brown with black tip; breast and belly russet with black streaks. Young covered with greyish-white down.

Australian kestrel (*Falco cenchroides*)

Birds also figure in the wedge-tailed eagle's diet, including crows, magpies and parrots.

Hunting methods are varied to suit the situation. The raptor may dive at a sharp angle from a height or perch patiently on a branch, awaiting the appearance of a potential victim.

The birds engage in spectacular nuptial displays. The male climbs to a considerable height and then falls, wings half-closed, pulling out of his dive at the last moment to soar aloft once more. Meanwhile the female flies in lazy circles, seemingly indifferent to his aerobatics and concerning herself with the care of the nest. Sometimes, however, she will turn on her back in mid-air and link claws with her suitor.

Both birds collaborate in building the eyrie, with the female doing most of the work. The nest is made of branches and may be 8 feet high and broad. Locations vary—high in the fork of a tree in sparse woodland but elsewhere among rocks or on the ground.

Mating does not necessarily follow immediately upon the courtship rites and may occur as much as three months afterwards, either in or around the eyrie. Some time later—often up to three weeks—the female lays from one to three eggs at four-day intervals. Each pair of eagles makes use of several eyries on the same hunting grounds and although nest building is an annual event the birds may return to the nest of the previous year if it has not been disturbed.

Incubation is the female's responsibility, the male providing food. After the birth he will continue to hunt for the family but this will normally consist of three members only, for as is so often the case with raptors, not more than one nestling is likely to survive. Within a month the first feathers sprout and a week later the eaglet makes its first attempts to tear up food supplied

by the parents. Plumage is complete at seven weeks and in its ninth or tenth week the fledgling will fly for the first time.

An even more characteristic raptor of the Australian desert is the grey falcon (*Falco hypoleucos*), slightly smaller than a peregrine falcon and resembling the African lanner falcon in its gliding skill and ability to catch prey on the ground. Flying in a leisurely manner with slow wing beats, this falcon hovers at low altitude, swooping on birds in mid-air and on lizards and small mammals. It consumes large quantities of insects as well. When not aloft it perches expectantly on the branch of a solitary tree but never ventures into the forest.

The grey falcon often makes use of a nest abandoned by a wedge-tailed eagle or fork-tailed kite. The female lays two to four eggs at some time between July and November, according to the particular region inhabited.

The Australian kestrel (*Falco cenchroides*) is often sighted in the interior, on cultivated land or even in large towns where it sites its nest on tall buildings. The gregarious habits of this species recall those of starlings in Europe. It hunts various small mammals and birds, and farmers tolerate it because its diet is comprised principally of insects. The female usually lays its four to five eggs in the hollow of a tree, a rock cleft or the abandoned nest of another species. When eight eggs are found in a nest they have probably been laid by two females. Incubation lasts 26 days and is carried out by the mother alone. For two weeks after hatching the chicks stay in the nest while either or both parents hunt for food. Around three weeks they embark on their first flight, but are not strong on the wing for a further twenty days. They fly well at ninety days after hatching.

The species is partially migratory. Some birds winter in the Celebes, Aru and Java; others are sedentary.

The wedge-tailed eagle (*left and centre*) hunts somewhat in the manner of vultures, the latter not being represented in Australia. This is evident in the form of the bill which is not typical of an eagle. Much persecuted, this raptor is becoming increasingly rare. The Australian kestrel (*right*), on the other hand, is found in most parts of Australia, often flocking in towns in a manner reminiscent of European starlings.

# CHAPTER 97

# The kangaroo: symbol of pride and prejudice

Australia is the land of the kangaroo. For centuries the Aboriginal tribes of the continent have depicted the animal in their rituals and totems and today its image appears on the national coat-of-arms. But although for foreigners the kangaroo symbolises Australia it is incorrect to speak of it as if it were a single species. In fact there are some 90 varieties, ranging from some which are no larger than rats to others that are as tall as humans. There are those which are essentially inhabitants of the inhospitable deserts of the interior and others that frequent the eastern rain forests. All of them, however, are vegetarians—apart from the musky rat-kangaroo (*Hypsiprymnodon moschatus*) which also feeds on insects.

In their isolation kangaroos were for long unknown to the nations of the West. In 1629 the *Batavia,* commanded by the Dutch navigator François Pelsaert, was wrecked off the west coast. A mutiny among the crew led to bitter fighting in the course of which more than a hundred men, women and children were slaughtered. This was followed by the trial and execution of the ring-leaders. Two of the crew were marooned, thus becoming the continent's earliest white settlers. Pelsaert was the first European to describe a kangaroo—actually a tammar (*Wallabia eugenii*). Rather vaguely, he spoke of a strange animal that appeared to be a species of cat, about the size of a hare and with a head resembling that of a civet! Inaccurate though it may have been, Pelsaert's report went almost unnoticed at the time.

Later in the 17th century other stories filtered through of curious cats, dogs, hares and sheep sighted in Australia; but it was not until 1770, as a result of an error of navigation, that the kangaroo was recognised and scientifically described by European naturalists. Captain Cook's *Endeavour* was trapped in coral reefs off the north-east coast and beached. The scientists made good

*Facing page:* Like the antelopes of Africa the Australian kangaroos are herbivores, well adapted to life on open plains.

use of their enforced stay on the mainland to study flora and fauna. Among the animals sighted were long-tailed grey creatures which moved about on their hind legs in a series of huge leaps. One of these strange beasts (the natives called them kangaroos) was eventually shot by one of *Endeavour*'s crew. Sketches and written descriptions later appeared in the published edition of this, the first of Captain Cook's voyages. Some years afterwards living specimens were sent home to England as presents for King George III and in 1794 kangaroos reproduced in captivity for the first time.

The animals commonly known as kangaroos are the largest members of the family Macropodidae; but strictly speaking the name applies to all the representatives of this family, both large and small. What they all have in common are hind legs that are much stronger and better developed than their forelegs, and a long, muscular tail. The second and third digits of the hind feet are linked with each other but the claws are separated, giving them a comb-like appearance – and in fact they are used for cleaning the pelage.

The typical marsupial pouch is well developed and is open to the exterior while the position of the body is often vertical. The central incisors, both in the upper and lower jaw, are very large.

The three characteristic kangaroo groups are regarded by some authors as subfamilies. They are the Hypsiprymnodontinae (musky rat-kangaroos), Potoroinae (rat kangaroos) and Macropodinae (kangaroos, tree kangaroos and wallabies).

The rat-kangaroos of the first and second groups were still present in large numbers about the middle of the 19th century but were to some extent decimated by the foxes imported at that period for the control of the rabbit population. Some varieties completely disappeared but the survivors now live in most biomes. The eight extant species belong to five genera—the broad-headed rat-kangaroos (*Potoroops*), the rufous rat-kangaroos (*Aepyprymnus*), the desert rat-kangaroos (*Caloprymnus*), the bettongs (*Bettongia*) and the long-nosed rat-kangaroos (*Potorous*).

According to Bernhard Grzimek the wallabies, kangaroos and tree kangaroos are divided into eleven genera but Erith and Calaby distinguish the following thirteen: hare-wallabies (*Lagorchestes*), banded hare-wallabies (*Lagostrophus*), nail-tailed wallabies (*Onychogalea*), rock wallabies (*Petrogale*), little rock wallabies (*Peradorcas*), tree-kangaroos (*Dendrolagus*), forest wallabies (*Dorcopsis*), New Guinea mountain wallabies (*Dorcopsulus*), scrub wallabies (*Thylogale*), short-tailed wallabies (*Setonix*), large wallabies (*Wallabia*), and large kangaroos (*Megaleia* and *Macropus*).

## The red kangaroo

A detailed animal distribution map of Australia will show that the large kangaroos fall into three well defined categories. The red kangaroos inhabit the heart of the continent while the grey kangaroos live in the comparatively wet regions of the south

*Facing page and above*: The smallest members of the Macropodidae are the rat-kangaroos. Shown here (*left to right*) are the long-nosed rat-kangaroo, the short-nosed rat-kangaroo and a pair of rufous rat-kangaroos. The last-named, although the largest of the group, do not normally measure more than 24 inches from tip of snout to tip of tail.

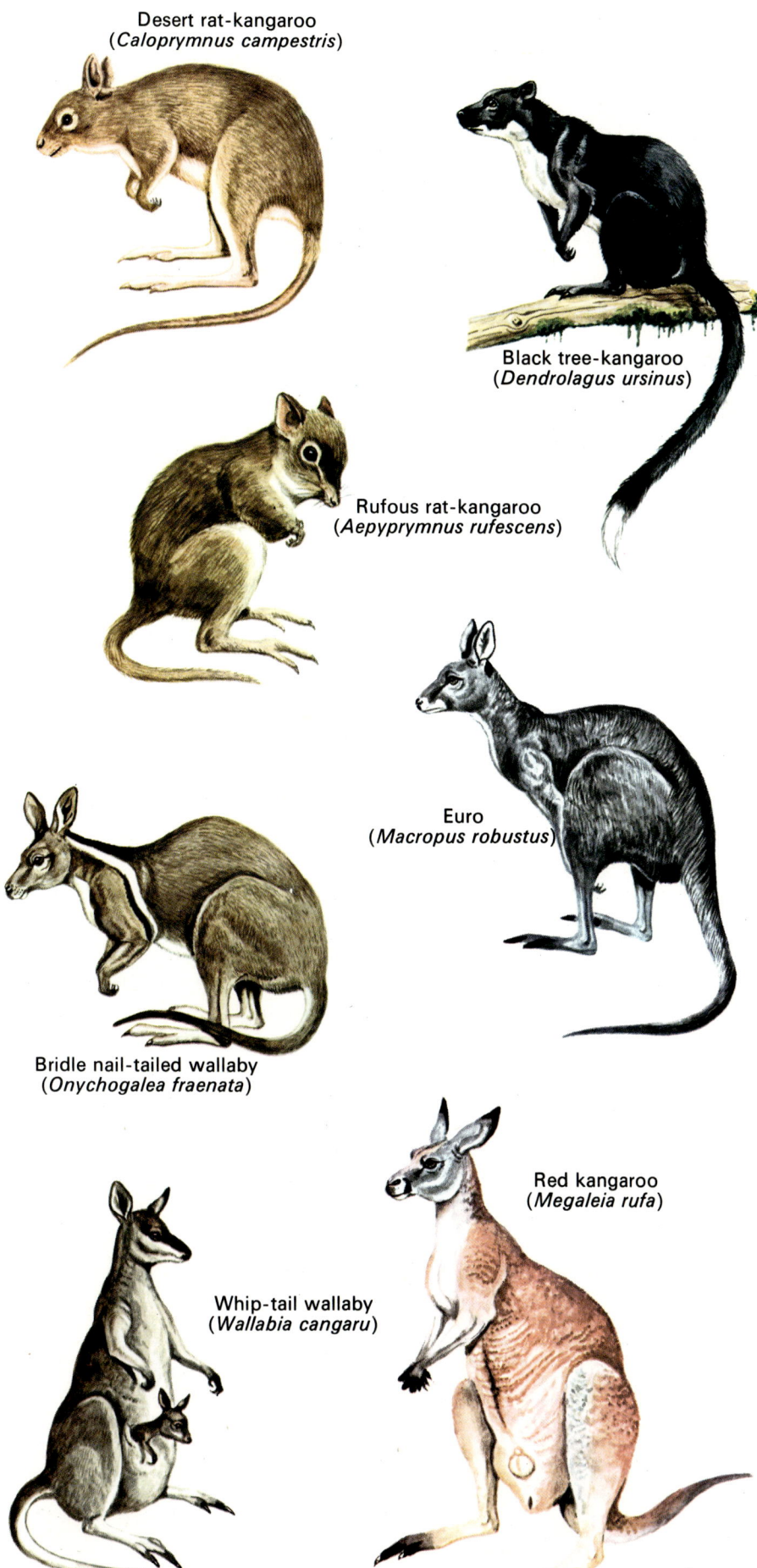

Desert rat-kangaroo
(*Caloprymnus campestris*)

Black tree-kangaroo
(*Dendrolagus ursinus*)

Rufous rat-kangaroo
(*Aepyprymnus rufescens*)

Euro
(*Macropus robustus*)

Bridle nail-tailed wallaby
(*Onychogalea fraenata*)

Whip-tail wallaby
(*Wallabia cangaru*)

Red kangaroo
(*Megaleia rufa*)

### LARGE KANGAROOS

Class: Mammalia
Order: Marsupialia
Family: Macropodidae
Diet: vegetation
Gestation: 30–40 days
Number of young: 1, rarely twins

Male larger and heavier than female. Powerful, well developed hind legs; forefeet small. Large ears. Strong, muscular tail. Marsupial pouch open at front.

#### RED KANGAROO
(*Megaleia rufa*)

Total length: 69–87 inches (175–220 cm)
Length of tail: 25½–41½ inches (65–105 cm)
Weight: 51–154 lb (23–70 kg)

Male's fur generally reddish, female's somewhat more bluish, although reverse may occur. Two annual moults, accompanied by colour changes. Top of muzzle hairy. Forehead whitish. Female's abdomen white, male's light brown.

#### GREY KANGAROO
(*Macropus giganteus*)

Total length: 63–94 inches (160–240 cm)
Length of tail: 29½–39½ inches (75–100 cm)

Fur colour varies with individual but no sexual dimorphism. Upper parts range from greyish-brown to greyish-red. Abdomen whitish. Lighter area around eyes. Muzzle hairier than in two other species mentioned here.

#### EURO
(*Macropus robustus*)

Total length: 53–90 inches (135–230 cm)
Length of tail: 23½–35½ inches (60–90 cm)

Colour generally dark greyish-brown, with slight variations according to subspecies. Abdomen pale. Digits and tip of tail very dark, sometimes black. Muzzle for most part hairless.

The quokka or short-tailed wallaby differs from other species in that its tail is not well muscled and is thus unsuitable for balancing the body. Once abundant, this species is nowadays restricted to some of the offshore islands in Western Australia. They prefer areas with thick grass and plant cover.

*Facing page*: Wallabies are similar in appearance to true kangaroos but rather smaller. The species shown here are the tammar (*above*), the ring-tailed rock wallaby (*below left*) and the red-legged pademelon (*below right*).

and east. As for wallaroos (the term used to describe marsupials that are midway in size between wallabies and kangaroos) they are scattered about almost everywhere. Some authors recognise only three species among these groups, *Megaleia rufa, Macropus giganteus* and *Macropus robustus;* but others distinguish a greater number. Thus Frith and Calaby make a clear distinction between the red kangaroo (*Megaleia rufa*) and the two species of grey kangaroo, the eastern (*Macropus giganteus*) and the western (*Macropus fuliginosus*). They also list three species of wallaroo, the black (*Macropus bernardus*), the euro (*Macropus robustus*) and the antelope kangaroo (*Macropus antilopinus*). Other authors use further methods of classifying these species.

The grey kangaroos inhabit regions with abundant vegetation (their distribution appears to coincide with areas affected by winter rains), the wallaroos are found in rocky and mountainous districts, and the red kangaroos on the open plains, the arid desert regions and the tree-covered steppes and savannahs.

After the arrival of the Europeans the habitat of the red kangaroo, somewhat surprisingly, was enlarged as a result of systematic land clearance and the catastrophic repercussions of the rabbit population explosion. On the other hand the programme of deforestation had precisely the opposite effect on the grey kangaroos and wallabies.

A series of articles published, among others, in the *New South Wales Gazette*, has described in detail how the European settlers progressively destroyed many of the continent's original biomes. When the first white men landed, for example, the Murrumbidgee plains were well wooded. No time was lost in cutting down trees and ruining the plant cover either by deliberate burning or indirectly by overgrazing. Throughout the country soil was laid bare and given over to crops and livestock. As a consequence the grey kangaroo vanished almost completely from these areas and its place was usurped by the red kangaroo.

The long-term effect of the white man's farming and stock-breeding activities proved, however, to be something less than beneficial to the red kangaroo population. According to Frith and Calaby the gradual decline of the species has been due to a variety of causes. They have been shot on sight by farmers, died from drinking poisoned water, starved as a result of land becoming so degraded that even the grass has disappeared, and been prevented from moving freely about by a network of fences and barbed wire.

The essential requirements of the red kangaroo are short grass, shelter and drinking water. Although the species' total area of distribution is vast (stretching uninterruptedly for something like a million square miles) the density per square mile fluctuates enormously, depending to what extent these vital needs can be satisfied. The animals are much more abundant in the areas where there is moderate rainfall and very sparsely represented in semi-desert and desert regions.

Even in a single area the density of the red kangaroo population will fluctuate seasonally; but although the animals lead a nomadic existence they cannot be said to have migratory habits. They simply wander from one pasture to another, often travel-

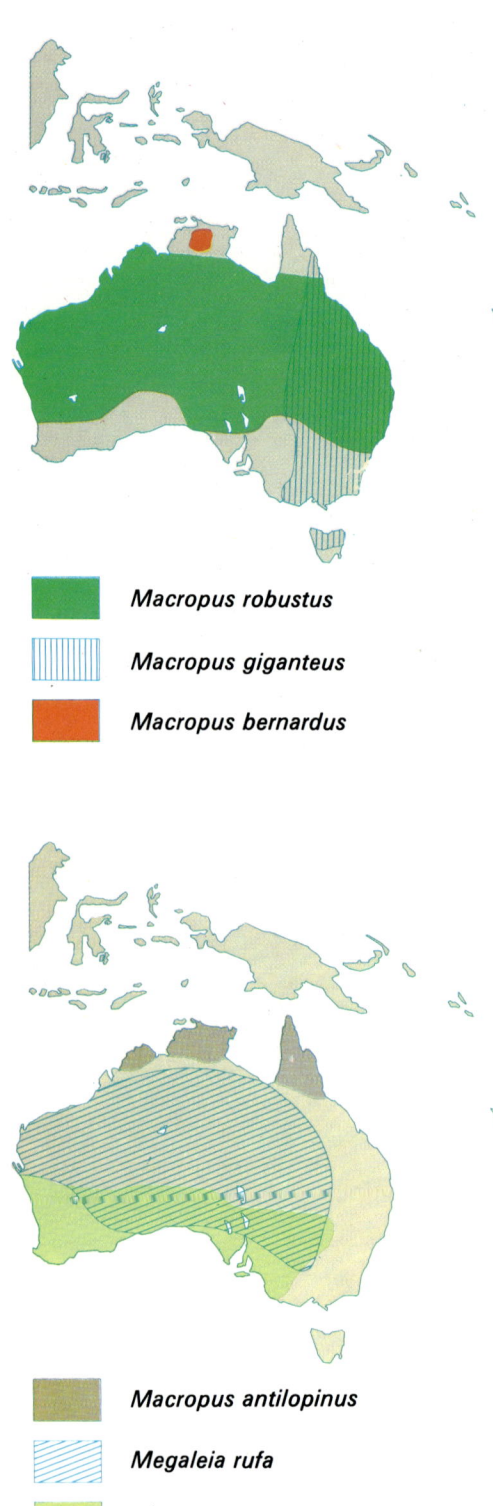

| | |
|---|---|
| ■ | *Macropus robustus* |
| ▥ | *Macropus giganteus* |
| ■ | *Macropus bernardus* |

| | |
|---|---|
| ■ | *Macropus antilopinus* |
| ▨ | *Megaleia rufa* |
| ■ | *Macropus fuliginosus* |

Geographical distribution, according to Frith and Calaby of the six species of large kangaroos.

ling enormous distances, like the gazelles of the African deserts, to reach those districts where it has rained recently and where fresh grass is plentiful. In former times hundreds and even thousands of red kangaroos would congregate on these green pastures, giving rise to two popular, but quite mistaken, assumptions. The first was that they constituted such a threat to domestic livestock that they had to be eliminated at all costs, the second that they were gregarious, living in large herds.

Although the red kangaroo needs to drink at regular intervals, zoologists have not thus far calculated exactly what quantity of water is required for survival. It is known, however, that the urine concentration of the species is higher than that of the wallaroo and more or less the same as that of typical desert mammals. This seems to suggest that the red kangaroo has evolved in an extremely arid environment.

## Kangaroo 'mobs'

Red kangaroos, timid by nature and active during the hours of darkness, are common enough in the wild but difficult to approach and study at close quarters in order to acquire scientific information. Their social behaviour has for long been a controversial subject. To ascertain what really goes on among the different members of a group the zoologist must not only make numerous, repeated observations but must also be able to distinguish one animal from another. This may seem obvious but in fact is no easy matter. Mistaken identity may be partly to blame for the stories that have been related of huge kangaroo herds consisting of hundreds of animals, dominated by a single large male who enjoys the favours of a harem of females. Recent investigations suggest that this is absurd.

Although large numbers of red kangaroos may be sighted from time to time grazing on a fairly small piece of land, these are simply 'aggregations', groups that have no social significance but which are formed solely because the district happens to provide adequate food and shelter. There is an analogy here with a phenomenon in the insect world, as when hundreds of butterflies congregate around water. The appearance of a horse, a car or an aeroplane is enough to cause panic in the group of kangaroos and split them into separate small units—known locally as 'mobs'—each of which makes off in a different direction.

Frith, who has made a population count from the air, has produced some interesting statistics relating to the size and composition of these mobs. After a survey of more than a thousand groups he published the following figures:–

| Number of individuals per group | 1-10 | 11-20 | 21-30 | 31-40 | 41-50 | 51-60 | 61-70 | 71-80 | 81-90 |
|---|---|---|---|---|---|---|---|---|---|
| Number of groups observed | 932 | 166 | 51 | 13 | 7 | 11 | 6 | 1 | 3 |

Frith's survey indicates that by far the majority of these red kangaroo mobs comprise from one to ten animals. In fact more

than half of them consist of from one to five individuals, the standard number being from two to four. Furthermore, when examining the composition of 193 groups Frith concluded that all of them contained males, but—what was particularly significant—in 119 cases there was only one adult male per group. Frith was fortunate enough to locate one group in which the largest male was an albino. He managed to slip a collar around this animal's neck and this helped him to draw certain tentative conclusions about the social behaviour of this particular mob, whose other members were three young males, three females and one baby. In the course of the observation another baby was born and although the group often joined other families it would invariably wander off on its own to sleep.

The most plausible conclusion to draw from these investigations is that each mob is a small family group. But a challenge to this theory is presented by the fact that solitary adults of both sexes are sometimes seen mingling indiscriminately with such groups, apparently encountering no opposition. Furthermore, nobody has yet proved that kangaroo behaviour is controlled by any system of rank that involves dominants and subordinates or that the strongest males possess a harem.

Ealey, who has carried out detailed marking experiments with groups of wallaroos, asserts that males and females are both solitary by nature and that they meet only at irregular intervals. Kirkpatrick, on the other hand, believes that among the eastern race of grey kangaroos the female is certainly the most important member of the group. He concludes that the only stable relationship between two individuals are those linking a mother and her baby.

The red kangaroo lives on the open plains and is much persecuted because of the contention of sheep farmers that it deprives their livestock of food. Recent studies have shown, however, that kangaroos and sheep, because they eat different types of grass, can share the same pastures without hardship on either side.

# A peaceful life

Although kangaroos are reputedly nocturnal animals their habits would appear to be dependent to some extent upon temperature as well as light. In the extremely hot parts of central Australia they venture out at sundown, making for their regular drinking sites, feeding at night and sleeping by day. But in areas where the climate is kinder the animals are quite likely to be active during the daytime as well.

Contrary to what might be believed of animals that are so well adapted to a dry, hot climate, the marsupials frequently seek shady spots during the day, stretching out beneath an acacia or beside a thorn bush, and sometimes even scooping a depression in the sand. As long as a mob remains in the same area each individual seems invariably to head for its own private sleeping site. Newsome underlines the importance of shade to the red kangaroos of the interior, and other zoologists have also seen the animals journeying several miles when the sun rises in order to find a shady refuge in the thickets.

Even when the animals are asleep there will always be one individual awake and vigilant. What is more, the kangaroos do not seem to sleep continuously. From time to time an animal will get to its feet, take a careful look all around, raise its head to sniff odours carried on the wind and then, if reassured, lie down once more. Sometimes it will bound away from the group and continue its sleep in another shady spot.

When the sentinel or any other alert member of the group senses danger, whether by night or day, it immediately signals the alarm to its companions. The animals keeping watch seem to

When moving about normally, as when feeding, the kangaroo makes short hops on all-fours with the tail as a fifth point of support (1). For travelling longer distances it propels itself forward with its hind legs only (2). When put to flight it puts more effort into the spring-like action of the hind legs, each single leap taking it a greater distance over the ground and into the air, the tail acting as a balance (3).

rely on keen hearing even more than on vision. In the opinion of Frith and Calaby the males are generally more nervous and uneasy than the females. In the event of imminent danger both sexes emit coughing cries which are instantly understood by all the kangaroos in the vicinity, whose response is to start drumming on the ground with their hind feet.

Kangaroos, when startled, make off with their characteristic bounding gait, hind legs stretched tautly like springs to provide powerful impetus, the tail acting as a balance. According to Troughton a large kangaroo is capable of jumping a distance of about 30 feet and a single leap may take it more than 10 feet up in the air. When alarmed the animal may travel at around 30 miles per hour but this speed can only be maintained for a few minutes at the risk of collapse from exhaustion. When moving about from place to place in a normally unhurried manner a kangaroo will cover 6–7 feet at most in a single hop, and its maximum speed is a mere 6–8 miles per hour.

While grazing the animal proceeds on all-fours in a crawl with forepaws and feet flat on the ground. From time to time it sits back, using the muscular tail as support. When upright and on the alert the kangaroo carries food to its mouth with its hands.

The tail is extremely important for locomotion, yet if by mischance the animal happens to be mutilated it can adapt itself to moving about on the hind legs alone. A grey kangaroo sighted near Canberra had lost practically all its tail but did not seem to be unduly handicapped, bounding around with agility and never falling behind its companions.

There seems to be little doubt that the typical hopping gait and upright stance of the kangaroo have been acquired as a result of the relatively flat and empty nature of the terrain where it lives. It is much easier to hear and smell at several feet rather than several inches above grass level. Just the same, many authors point out that the hopping movement of the kangaroo

Although kangaroos are usually regarded as nocturnal animals, their activities are to some extent controlled by temperature as well as light. When it is not too hot they will venture out by day but whenever the heat is intense they take their ease in whatever shade is available.

The red kangaroo can reach a speed of about 30 miles per hour but only over a comparatively short distance.

is a far less efficient way of getting around than the all-fours bounding gait of savannah ungulates, suggesting that the former have adapted to environmental conditions less successfully, for example, than antelopes.

As a rule kangaroos are unafraid of water and will plunge into streams and rivers if pursued by dogs. They devote much time to hygiene, scratching and cleaning their fur with the comb-like toes of their hind feet or with the five fingers of the hands.

## Breeding habits of the kangaroo

Female kangaroos living in the driest regions of Australia are comparatively late in reaching sexual maturity, at about the age of three years. Those that inhabit wet regions (the term is of course relative) are mature at one and a half years. The males seem to be less affected by local climatic conditions. Wherever they live they are capable of procreating at the age of about twenty months. It sometimes happens, however (and this applies both to males and females) that prolonged drought, resulting in shortage of food, may delay sexual maturation.

The male recognises a female on heat by her distinctive odour and then proceeds to court her, following her around and making soft grunting noises. Now and then he attempts to catch hold of her tail, which does not appear to please her greatly, for she bounds off, uttering grunting noises. Far from intimidating her partner these sounds seem to excite him all the more. Eventually she ceases to resist and adopts a submissive crouching posture, hands on the ground. The male mounts her and the sexual act continues for perhaps fifteen or twenty minutes.

Although not invariably caused by the presence of a female, brief clashes between males are frequently observed. A kangaroo will try to seize a rival by the arms and to kick him in the belly. Despite a number of reports of fierce and bloodthirsty battles

According to Troughton, the tremendous leaps of a red kangaroo may measure up to 30 feet in length and 10 feet in height.

The male red kangaroo recognises a female on heat by her odour and tries to catch hold of her by the tail. When ready for mating she leans her forefeet on the ground and permits him to mount her.

When two male kangaroos confront each other they rear up on their hind legs. Stories of them 'boxing' are, however, unfounded for they use their arms only to prevent the opponent from moving.

the only recorded incident of a fight between two male kangaroos in which the loser died of such injuries occurred in captivity, the space in which the contest took place being so confined that the wounded animal had no chance to escape.

Aggressive behaviour of this nature is also displayed when a kangaroo is cornered by an enemy, especially a man or a dog. But there is no truth in the hoary fable that kangaroos defend themselves by 'boxing'. If the arms are used the purpose is only to immobilise the opponent in order to deliver a hefty kick with the hind legs.

The gestation period is thirty or forty days, according to the species. Thus in the case of the red kangaroo it is thirty-three days. There have been a number of reports of females that have given birth several months after their last recorded mating. In 1912, for example, Carson stated that he had found a baby kangaroo in its mother's pouch 332 days after the death of the only adult male sighted in the area. It was only some fifty years later that it was definitely established that, as in the case of roe deer, badgers and martens, the blastocyst is sometimes delayed in becoming fixed to the walls of the uterus—another example of the phenomenon known as delayed implantation. The blastocyst is a very early stage in embryonic development.

Naturalists now know that the female kangaroo is capable of mating again soon after the birth of a baby, but that as a result of delayed implantation the development of the embryo is arrested, so that a second baby is born only after the previous one has completed its growth and left the pouch. Thus a female may be carrying an embryo at the same time as she is caring for and feeding another.

There is no overt sign of her being pregnant. The first hint that a female may shortly be due to give birth is when she begins cleaning her pouch. This is open at the front and contains four teats inside, at least two of which provide a continuous supply of milk.

A day or so before the birth takes place the female lies on her back and inserts her head into the marsupium to clean it. The cleaning process becomes more nervous and vigorous as the time for the birth draws nearer. In addition she will lick her uro-genital region, her chest, the soles of her feet and the base of the tail.

## Birth adventure of a kangaroo

When in the 19th century John Henderson saw for the first time a baby kangaroo in its mother's pouch he tried to explain it by putting forward a rather fanciful theory on the subject of marsupial reproduction. In his view the female's pouch was an open womb, inside which gestation proceeded in exactly the same fashion as with placental mammals. He stated that there were two teats inside the marsupium and that the mother gave birth to a single baby, the foetus making its appearance at the tip of the teat.

We now know that Henderson's ingenious theory has no foundation whatsoever, based as it was on the beliefs of the

Aborigines. Yet the true facts, as scientifically verified, are hardly less astonishing. At birth the baby kangaroo is minute, weighing barely one gramme and measuring approximately one inch. Its tiny body is entirely without hair, the eyes and ears are not completely formed and the hind legs are no longer than the front ones. In fact at this stage it resembles in every way an embryo of a placental mammal at the very beginning of its development. Nevertheless, secure and warm inside the marsupial pouch, this unformed creature attaches itself to the mother's teats, exercising such strength that if it is forcibly separated a few drops of blood may fall from the place where it has been suckling. The intriguing question, of course, is how did the newborn baby manage to get into the pouch?

Although every birth is a great adventure, that of the baby kangaroo is even more so. Tiny and apparently incapable of any kind of co-ordinated movement, the undeveloped baby has to find its way into the maternal pouch and take refuge there the moment it is born. For some time experts thought that the mother took it in her mouth and deposited it in the pouch. But the real answer to the mystery is now available, since the filming by Sharman and Frith of the birth of a red kangaroo at Canberra.

Guided by its sense of smell—for the other sensory organs are not thus far functional—the little creature, still attached by a thread of umbilical cord, undertakes the most important journey of its life, from the uro-genital tract of the mother to her teats. It takes about three minutes to move up to the mouth of the marsupium, clutching tightly to the hairs of its mother's abdomen. Once it has found the way into the pouch it spends about a quarter of an hour trying to attach itself to a teat. Having done so it will not relinquish its hold for some six months. When at the end of that period it leaves the pouch it can really be said to be born.

Sharman and Pilton point out that the newborn red kangaroo possesses five fingers on either hand which are already well developed and which have strong claws; but there is as yet no sign of the four toes of the hind feet. The baby therefore makes its way up the abdomen of the mother by using only its forelimbs, gripping so firmly that a considerable effort is needed for anyone to prise it loose.

The red kangaroo, like the emu, is an inhabitant of arid regions, provided there is some water in the area to quench thirst.

A male kangaroo may try to intimidate a rival by standing upright on the hind legs and balancing on his tail. If the other animal responds in the same manner a battle may ensue in which each tries to land kicks in the opponent's abdomen.

*Facing page:* The birth of a baby kangaroo is an amazing adventure. The first sign that the female is shortly due to give birth is when she begins cleaning her pouch. When it emerges from the cloaca the inch-long baby is naked and its eyes and ears are incompletely formed. Despite these handicaps it makes its way up to the marsupium, clinging with the claws of the forefeet to the hair of the mother's abdomen, a journey lasting several minutes. Having reached the pouch it attaches itself firmly to a teat and does not let go for another six months.

The mother regularly licks her pouch clean as the moment of birth approaches (1). The arrow shows the path taken by the newborn baby from the uro-genital opening to the marsupium (2).

It appears that on some occasions a mother will assist her baby to reach the pouch; but in an experiment carried out with an anaesthetised female it was found that the baby was not only born normally but managed to find its way to the teat without any difficulty. At one time it was thought that the mother traced out a path in her fur for the baby, but apparently this is not the case, the error having arisen from the fact that the animal licks her belly both before and after the birth.

A kangaroo almost always gives birth to a single baby. At London Zoo, nevertheless, eleven sets of twins and one pair of triplets were born to kangaroos or wallabies out of a total of 119 births. In one multiple birth at Philadelphia Zoo the mother of twins was seen to toss one baby out of the pouch so as to rear the other by itself.

## In and out of the pouch

Following the birth of her baby the mother continues to lick and clean herself for the best part of an hour. Then she seems to forget her offspring completely although she cleans the pouch several times a day while the baby is growing. As long as the latter is naked she demonstrates no affection for it but the situation changes when, after about 190 days, the baby kangaroo, now covered with hair, leaves the pouch for the first time. Initially this will only be for a couple of minutes but later the baby will stay out for progressively longer periods. When the group moves from place to place or is in some way threatened the mother summons her youngster (which hops towards her) and inclines her body forward to help it climb back into the pouch. The little animal enters head-first and once inside performs a complete somersault so that it ends curled up, with head, hind feet and tip of tail just protruding.

If the young kangaroo strays too far from its mother's side it utters a series of loud yelps—the same noises in fact that it has been making for some time inside the pouch but without provoking any interest on the mother's part. Now, however, the sounds elicit a response not only from the mother but from all the other females in the vicinity. Conversely, in the event of the mother losing track of her offspring, she will search for it in every direction, to the accompaniment of her own questioning cries.

At the age of about seven and a half months the young red kangaroo weighs 7–9 lb and is big enough to leave the pouch. The young grey kangaroo stands on its own feet about a month earlier and a young wallaroo at about seven months. When the mother gives birth to another baby the firstborn continues to follow her around in order to suckle but confines itself to placing only its head into the pouch.

Having to transport a youngster weighing 4–6 lb may be just as inconvenient and dangerous for a kangaroo as it is for a placental mammal to carry a baby inside her body. The additional weight handicaps free movement and slows down flight. But whereas a gazelle, for example, can do nothing about excess weight when pursued by a leopard, the kangaroo can get rid of her load quite

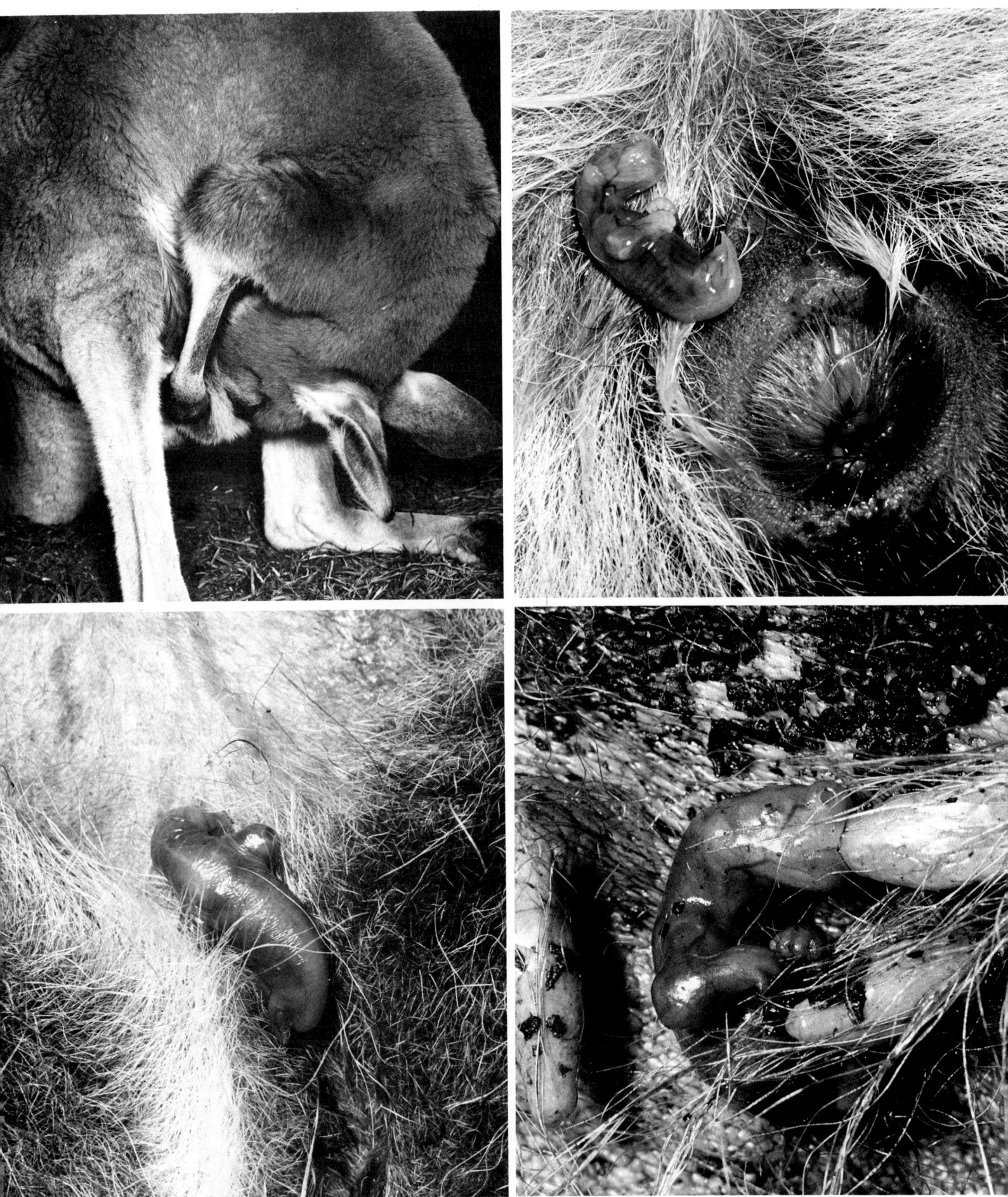

When the young kangaroo or joey is big enough to wander about on its own it returns regularly to the pouch, first inserting its head and then executing a somersault so that the head, hind legs and tip of tail protrude.

*Facing page*: A kangaroo mother may find herself caring for two youngsters of different ages at the same time. The younger will be carried about in the pouch while the elder, not yet weaned, suckles by introducing only its head into the marsupium.

easily when hunters and their dogs get too close, simply by dropping the baby out of the pouch. Although legends cite cases of mother kangaroos returning to look for the babies they have abandoned in this manner, the grim reality is that once having resorted to such extremes the female has practically no way of telling exactly where she may have dropped the joey. If the latter still lacks its covering of fur it cannot possibly survive out of the pouch; and even when its pelage is complete the young kangaroo is no less vulnerable. Solitary youngsters are never encountered for the simple reason that once on their own they are picked off by predators. It is true that in captivity female kangaroos have been known to adopt orphans; but in the view of Frith and Calaby there is nothing to indicate that this will happen in the wild.

Quite often the joey attempts to climb back into its mother's pouch when it is too large to be protected there any longer. In such a case the mother prevents it doing so by taking hold of the youngster by the front legs. Sharman and Pilton describe how one female managed to convince her baby that it was too old to be carried around in the pouch by repeatedly tumbling it to the ground. In the normal way she will suckle her youngster until the end of its first year.

## Sheep versus kangaroo

Australia is famed for its sheep farming and for this reason bitter antipathy has arisen between the farmers and those who actively campaign for the protection of the kangaroo. The sheep farmers are convinced that kangaroos are detrimental to their flocks by grazing on the same pastures. This is certainly a delicate and highly controversial problem which revolves around the question of the comparative feeding habits of the species concerned.

A certain amount of green vegetation is usually to be found in the stomach of a kangaroo. Detailed examination of this indicates, however, that it consists almost entirely of short, fresh grass to the exclusion of tall grass and straw (these being consumed only in the event of acute food shortage). Even in very dry regions the kangaroo will concentrate only on the most succulent grass whereas a sheep will consume almost any type of vegetation. Nevertheless, despite the fact that the gestation and suckling periods of sheep are longer and that twin births are more frequent, kangaroos reproduce more rapidly in these arid regions, thanks to the afore-mentioned phenomenon of delayed implantation. Thus if there is prolonged drought at a time when sheep and kangaroos happen to be giving birth, the former have to await a new rutting period before making good their numbers whereas the latter lose no time in having other babies. The consequences are evident, for example, around Pilbara in north-west Australia where, according to Grzimek, the number of sheep has decreased by half in the last twenty-five years. More than a dozen farms that once contained eight million sheep between them have had to be abandoned, while the local population of wallabies and other small kangaroos has multiplied.

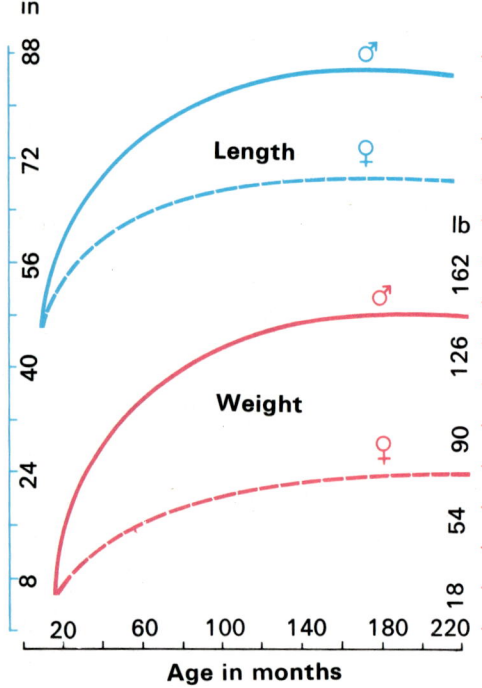

Graphs showing increase in length and weight of a red kangaroo at different ages (according to Frith and Calaby).

*Facing page:* The dingo, Australia's wild dog, is one of the principal predators of the continent, which probably arrived with the Aborigines from Asia. Prior to the appearance of the white man it was the chief enemy of the kangaroo.

Frith and Calaby point out that the arguments both for and against the kangaroo still rest on inadequate scientific information about their feeding habits. But sufficient work has been done on the subject to show that although sheep and kangaroos sometimes eat the same type of vegetation the main constituents of their respective diets are quite different and that consequently they can live together on the same pastures. Griffiths and Barker, among other authors, make the point that among the short grasses *Amphipogon cariginus* is eaten by sheep but shunned by kangaroos. Conversely, the so-called kangaroo grass (*Themeda australis*), much favoured by the marsupials, is normally passed over by sheep.

Both types of animal are able to live under extremely arduous conditions. Sheep are capable of withstanding an increase in their own body temperature without it affecting their normal activities. If no water is available they can shed up to a quarter of their body weight without harm. Kangaroos are equally well protected against intense heat. Like dogs, they pant to keep themselves cool and continually lick their hands, arms and chest, thus lowering the body temperature as the saliva evaporates. Furthermore, they often dig holes in the dried-up beds of rivers, up to 3 feet in depth, in order to reach the water below the surface, just as elephants do in Africa. Needless to say, by opening up these water holes they perform a valuable service for other animals.

Ealey has shown how even in arid districts where men have sunk wells or tapped freshwater springs wallaroos seldom use them for drinking. Studies on nutrition carried out at the University of Western Australia have also indicated that the less an animal drinks the better it can assimilate nutritious substances from desert plants that are poor in protein. Dr Main has demonstrated that the tammar (*Wallabia eugenii*) which lives in south-west Australia and on some offshore islands (places where it may not rain for months) can survive by drinking sea water.

Another important discovery is that the flesh of a kangaroo contains a higher proportion of muscle (that is, edible protein) than that of most forms of livestock. A dead kangaroo is 52 per cent muscle whereas the figures for pigs and sheep are 32 per cent and 27 per cent respectively. Yet although kangaroo meat, properly prepared, can be excellent for human consumption it has acquired a world-wide reputation as being fit only for pet food and as such is exported in quantity.

Only a few of the smaller species of kangaroo are threatened with extinction as a result of hunting and habitat destruction. Other species, particularly the largest, are still found in sufficient numbers in certain parts of the continent to make it feasible for them to be culled in a sensible manner so that the meat could be used for feeding people in famine areas.

## The kangaroos' natural enemies

Before man and his animals reached Australia the kangaroo had little to fear from natural predators for even the Tasmanian wolf was not always fast enough to catch it. In those far-off days, apart

from adverse climatic conditions and shortage of food, the principal cause of declining populations was disease, mainly affecting young animals.

Man's arrival radically altered the situation. It was not the primitive hunters who were chiefly responsible, for their weapons were rudimentary and inaccurate; but the dogs which accompanied them proved to be formidable predators.

Many authors maintain that the Australian wild dog, the dingo (*Canis dingo*), is virtually the same species that arrived with the distant ancestors of the Aborigines and which later returned to the wild. Supporting this theory that the dingo is descended from a race of domestic dog is the fact that the animal exists in a variety of colours, although normally it is reddish. Professor Macintosh, however, emphasises a number of features that distinguish the dingo from the domestic dog, ranging from sexual habits to the protein content of blood serum. Taking all the arguments into consideration, it still seems probable that the dingo did arrive with primitive man from Asia, although the date may be uncertain.

Whatever its origin the dingo is one of the most important predators in Australia. Despite the fact that systematic attempts have been made to exterminate the species, its numbers are still sufficiently large for it to exert some influence on the marsupial population. Frith and Calaby state quite positively (although they have not observed it directly) that dingoes hunt kangaroos in packs and end up killing them; certainly they have been seen in groups feeding on corpses of red kangaroos. It is possible that these are in fact family units and that the dingoes pursue their prey in much the same manner as African hunting dogs and Asiatic dholes.

Like many other Canidae, the male dingoes mark their territory with urine. The animals mate in winter and the puppies are born in spring after nine weeks' gestation. A litter may comprise from five to seven or up to nine babies which are suckled for two months and protected by the parents in the burrow. For the first year, and sometimes longer, the puppies depend on the adults who instruct them in hunting.

Although the dingo is the most powerful ground predator on the continent it too has enemies. In addition to human persecution the dingoes in tropical regions are killed by crocodiles and snakes. But probably the greatest danger comes from the air in the shape of the wedge-tailed eagle. Ornithologists have watched the eagles hunting in pairs and killing dingoes in this manner. Although this does not necessarily happen very often there is no doubt that puppies as well as old and sick individuals are habitually preyed upon by this formidable raptor.

It is likely that the European fox, introduced to Australia in the expectation of controlling the rabbit population explosion, also kills young kangaroos should they stray too far from their mothers; but examination of the stomach contents of the species indicates that this is a rare event. On the other hand, the populations of rat-kangaroos and small wallabies have been adversely affected by the wholesale importation of foxes, to the point of bringing them to the brink of extermination.

The wedge-tailed eagle also preys on kangaroos from time to time. Leopold and Wolfe in their examination of the contents of several eyries, came across a large number of rabbit bones but also found many remains of kangaroos, estimating their ages at between seven to ten months – precisely the period when the youngsters would normally leave the pouch. It is difficult, just the same, to make a final pronouncement on this matter since, given the scavenging habits of this raptor, one cannot determine whether in such circumstances the animals would have been killed directly or whether they might have died of other causes and simply been picked up as carrion.

As happens with many other species, kangaroos are their own worst enemies when numbers become too great. Thus, as Dr Main has described in the case of the tammar in its island habitats, overpopulation and overgrazing leads these animals, which live in runways concealed in tall grass, to defend their territory fiercely against intruders, so much so that many of them never acquire a patch of their own and are consequently at the mercy of predators. Exposed on open terrain, such animals do not survive for very long.

Although it has not been observed at first hand, dingoes are believed to hunt kangaroos, this assumption being based on the fact that small groups of these wild dogs have been seen feeding on their dead bodies.

# CHAPTER 98

# Mammals of the Australian woods and forests

Australia is predominantly an arid, treeless land. Only around the fringes of the red desert which covers the interior are there narrow zones of wood and forest, especially in the east and south, making up only about 2 per cent of the continent's total area. The commonest tree is the eucalyptus or gum tree, of which there are some 600 species.

This varied and discontinuous belt of trees changes its appearance in relation to the different climatic conditions that prevail locally. In the north the monsoon rains reach a climax during the Australian spring and summer, from November to April, and there is fairly heavy rainfall on the coasts. But in the south the chief climatic influence comes from the Antarctic. Only a comparatively small area is affected by rain, which is at its heaviest in winter. The varying rainfall pattern naturally determines plant growth in different parts of Australia.

The north-east regions of the country (coastal Queensland and northern districts of New South Wales) are hotter and wetter than elsewhere. The eastern highlands retain a large amount of moisture and this has given rise in the north to an area of tropical mountain forest. But although there are only small climatic fluctuations along the coast, the rain forest does not stretch continuously because of fires and land clearance schemes. Nowadays there are only scattered zones, some of them quite large, of mixed vegetation, but the rest of the former forest has been given over to cultivation, dairy-farming and the growing of sugar cane.

In what is left of the primitive Australian jungle the traveller can still lose his direction in the dense undergrowth; the botanist may still marvel at the delightful colours and remarkable shapes of innumerable species of orchids; and the zoologist

*Facing page:* Although Australia is for the most part arid and hot there are belts of forest, particularly in coastal regions, some of them quite extensive. Heavy rainfall in some areas makes for a varied and abundant flora.

will find rich reward in the study of several highly interesting groups of animals.

It is a strange paradox that, at first glance, it seems to be various species of conifers, traditionally associated with cold and temperate climes, which dominate the coastal forest scene. Bunya and hoop pines (*Araucaria*) and *Agathis* tower beyond 150 feet, their summits forming an impenetrable green canopy which prevents the sun getting through to the grass and shrubs at the lowest levels.

The belt of open tropical woodland which partially surrounds the rain forest is much more extensive. Here the slender eucalyptus and bulging bottle tree are the main features of the landscape, mingling, on the fringes of the savannah, with acacia. Herbaceous plants spring up among the trees in the rainy season, only to wither and vanish with the return of drought.

In the south-east abundant rainfall and rather lower temperatures have stimulated the growth of dense woods of eucalyptus, some species of which, notably *Eucalyptus gigantea* and the mountain ash, *Eucalyptus regnans*, reach enormous heights, sometimes over 300 feet. In the more southerly districts the gum trees are interspersed with beech (*Nothofagus*). In Tasmania there are extensive tracts of beech woods.

In the extreme south-west, as well as in the Adelaide region, the climate is of the Mediterranean type. Towards the centre of the continent rainfall becomes progressively lighter and more irregular. Here the woods are very scattered, comprised principally of eucalyptus, together with she-oaks (*Casuarina*) and acacias. The main characteristic of the open plains is, however, a wide variety of grasses. Most of the plant growth is clustered along creeks and rivers. One typical tree found in the vicinity of water is the river red gum (*Eucalyptus rostrata*), which is in fact the most widely distributed of all Australian eucalyptus species. Despite its name this tree is well adapted to changing temperatures by reason of possessing two types of root, one set almost at ground level (thus absorbing surface moisture), the other much deeper, tapping underground water. Taking the various wooded areas as a whole, therefore, it is undoubtedly the hundreds of forms of the genus *Eucalyptus* that dominate the scene, thanks to their remarkable qualities of adaptation.

It is interesting to note that man's handling of the eucalyptus has had a profound and adverse impact on forest development wherever it has been introduced, just as his importation of placental mammals has decimated the marsupial population. But in this case the movement was in the opposite direction for the eucalyptus has been exported to many foreign countries. In a Mediterranean climate, for example, the eucalyptus grows at an astonishing rate so that endemic species such as beech, oak and the like are unable to contend with it for space. Furthermore, no plants will grow in the shade of this tree and consequently the woods artificially formed by planting various species of *Eucalyptus* are unsuitable for local fauna. Thus the apparently innocent introduction of trees from overseas has upset the ecological balance of many regions. Hopefully it may serve as a warning against similar experiments in the future.

There are some 600 species of eucalyptus in Australia, well adapted to changing climatic conditions. This is the snow gum (*Eucalyptus niphophila*) which withstands extreme cold in high mountain regions.

## The koala: specialised inhabitant of the eucalyptus forests

The koala (*Phascolarctos cinereus*), a small tree-dwelling marsupial, is probably the most popular of all Australian mammals, even though it is not found in many of the world's zoological gardens. The earliest report of this delightful animal's existence came in 1798 when John Prince, the young secretary to the Governor of the time, mentioned an endemic species known to the Aborigines as *cullwine,* which seemed to him to resemble the sloths of Central and South America.

Measuring between 2–2½ feet, the koala, its stocky, tailless body covered with thick grey woolly fur, tinged with brown, is a natural charmer. With its round shaggy ears, sleepy eyes and prominent black snout, it looks like a bear cub. The first digit of each hind foot is opposable to the other four, and on the forefeet the first two digits (thumb and index finger) are opposable to the other three, so that each hand virtually possesses two thumbs. By reason of this finger and toe pattern, and the fact that they are all provided with sharp, curved claws, the koala can grip firmly the trunks and branches of the gum trees in which it lives, moving about with complete confidence and great agility.

Of all living marsupials the koala has by far the most specialised

The tree growth of the Australian savannah is dense in places but scattered. This picture, taken in Queensland, shows damage caused to trees by domestic livestock, one of several reasons for the gradual degradation of plant cover.

Geographical distribution of the koala.

**KOALA**
(*Phascolarctos cinereus*)

Class: Mammalia
Order: Marsupialia
Family: Phascolarctidae
Length of head and body: 24–32 inches (60–80 cm)
Tail: vestigial, hardly visible
Weight: 9–33 lb (4–15 kg)
Diet: eucalyptus leaves and shoots
Gestation: 25–30 days
Number of young: one

Large head and muzzle, shaggy ears. Thick ash-grey fur. Feet adapted for climbing; strong claws. Thumb and index finger of hands opposable to other three; big toe of hind feet likewise opposable to other four. Caecum measures 6–8 feet. Marsupium, with teats inside, opens towards rear. Baby, weighing barely ¼ ounce at birth, makes its way into the pouch, remaining there for six months suckling but leaving it periodically to imbibe viscous green fluid (predigested vegetation) flowing from mother's anus. Dependent on mother until one year old, the young koala is sexually mature at three or four years of age.

*Facing page:* The koala, popularly known as the Australian teddy bear, lives in gum trees, feeding exclusively on their leaves and shoots.

vegetarian diet, for its food is almost exclusively the leaves and shoots of one type of tree, the eucalyptus. This is narrowed down still further, for although there are hundreds of *Eucalyptus* species there are, according to Grzimek, only twenty or so that the koala finds acceptable, and out of these five alone stand out as being particularly favoured. Among this small group a clear first choice is the manna gum (*Eucalyptus viminalis*), followed by the striped eucalyptus (*Eucalyptus maculata*) and the forest red gum (*Eucalyptus tereticornis*). But in addition to being highly selective where species are concerned the koala, at least in the wild, is equally particular about which leaves of the tree to eat. This fact first came to light in the course of an investigation by Ambrose Pratt into the alarmingly high seasonal death rate of captive koalas in Melbourne Zoo. The mysterious illness to which the animals regularly succumbed resisted all forms of medical treatment, and it was only by accident that the zoologist happened to examine the chemical composition of the eucalyptus foliage on which they were fed. To his astonishment analysis of this vegetation revealed the presence of large quantities of prussic acid in the leaves and tender shoots of the species concerned.

These very disturbing discoveries led to more detailed study of the koala's standard diet, which showed quite clearly that in the case of most *Eucalyptus* species, and particularly *Eucalyptus viminalis*, the young shoots and leaves contain an unusually high degree of toxicity, above all in winter. This is not of great significance in the wild, for the marsupials–by a combination of experience and maternal instruction–consume only those leaves and shoots that are low in toxicity. In captivity, however, the animals were given the diet which was assumed to be the best for them, namely the softest and most succulent parts of the foliage. Unknown to anyone at the time, these happened to contain particularly high levels of poisonous acid. So the koalas were dying simply because they were not being given the absolutely free choice of food that they would have enjoyed in the wild.

Not every koala, however, possesses precisely identical food habits. Much depends on the area inhabited. Although the koala population is concentrated in the eastern coastal forest strip, there are a number of different races to be found, each with slightly differing tastes. Zoologists have observed, for example, that the animals living in Victoria refuse to touch greyish or bluish eucalyptus leaves, which are greatly to the liking of koalas inhabiting forests in Queensland. Thus animals from different regions clearly have an instinctive preference for particular local gum tree species. This poses a problem where allocation of land for nature reserves for koalas is concerned.

The fact that the koala insists on such a highly specialised diet creates obvious problems when it comes to keeping the animal in captivity, and it is principally for this reason that, for all its popularity, it is comparatively seldom to be seen in overseas zoos. Quite apart from having to employ an experienced dietician to ensure that the koala receives exactly the correct type and amount of food, it is an extremely expensive business

Opposable digits on hands and feet enable the koala to retain a firm grasp of the trunks and branches of the trees which are its home. Only a handful of the many species of eucalyptus are frequented, these being the ones which provide the right type of foliage for eating.

to arrange for a continuous supply of fresh eucalyptus foliage, bearing in mind that one animal needs as much as 2–3 lb of food every day.

Like all exclusively plant-eating animals the koala's digestive apparatus is adapted in such a way that maximum nutritional value is extracted from the food consumed. Modification of the intestine, in the form of an extraordinarily long caecum, measuring some 6–8 feet, makes it possible for the animal to digest such a large quantity of foliage. This facilitates prolonged maceration of all ingested food, assisting the fermenting action which ensures that everything is properly digested.

## Special food for the young koala

Following the normal suckling period there is apparently a stage of feeding in which the baby koala is prepared for its adult diet of vegetation. During this time the baby, while still developing inside the marsupium, takes in a kind of semi-liquid pap produced by the mother. This substance and the amazing way in which it is administered was first described in 1933 by the naturalist Keith Minchin. Whilst studying the behaviour of a group of koalas he was surprised to see a baby poke its tiny head out of the mother's pouch (which in this species, as in the wombat, opens backwards) and, by clinging with the claws of the hands, make its way down her abdomen to the anal opening. Thrusting its head inside, the baby then proceeded to suck a pale green secretion flowing from the mother's anus. The fact that the female had that day produced faeces of the normal colour and consistency indicated beyond dispute that the fluid on which the baby was feeding had no connection with the excreta and was in fact something quite special, apparently predigested material transformed in the caecum into easily assimilated pap.

Although it is not known exactly how this substance is formed it can be stated with certainty that it bridges the gap between the milk and vegetational diet. The digestive mechanism of the young koala has to be prepared for the new type of food it will eat when it becomes independent of its mother. Before feeding on leaves there must be present in the animal's intestines the protozoa and bacteria that by their fermenting action break up the material and make it assimilable. The anal secretion of the mother evidently contains the bacteria that have to be introduced into the digestive system of her offspring.

The fact that the marsupium of the adult female koala opens towards the rear is of some assistance to the baby as it crawls along the abdomen to the anal orifice in its quest for predigested pap; but in other respects the structure of the pouch is hardly an advantage for a tree-dwelling animal. When the mother is continually climbing up and down the trunk or scrambling from branch to branch the youngster has to maintain a very tight grip on her body in order not to fall. There would be no danger of this happening were the pouch to open towards the front, as is the case with the various kangaroo species.

The young koala lives on this special pap diet for about a month, after which it is adequately prepared to feed on tough eucalyptus leaves. But although it now feeds independently it is not yet of an age to leave its mother and continues to travel astride her back for several months to come without her care and protection.

## The imperilled koala

There are few carnivores in Australia and certainly no large tree-dwelling predators to threaten the tranquil existence of the koala. In former years, therefore, the species ranged far and wide through the eucalyptus forests, their peace interrupted only by the occasional hunting forays of the Aborigines, and their numbers periodically reduced only by disease. But as happened with most of the other indigenous Australian animals, this carefree life ended with the arrival of the white man. At first hunting was a matter of subsistence and the little marsupial was killed in order to provide the early settlers with food and certain articles of clothing. But with the development of firearms hunting the koala grew to be a flourishing and profitable

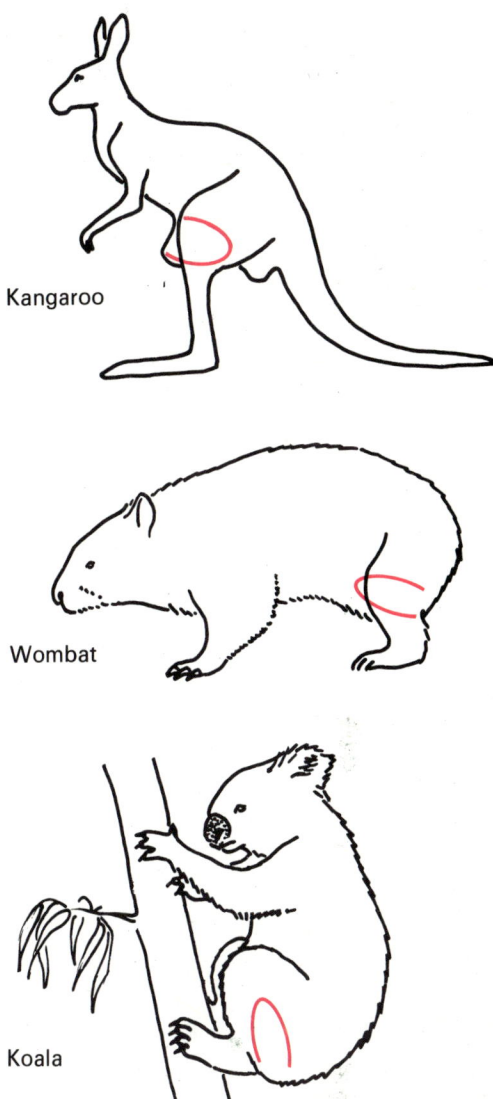

Kangaroo

Wombat

Koala

Whereas the marsupium of the kangaroo family opens towards the front, enabling the baby to climb in and out without much difficulty, in the case of the wombat and the koala it opens to the rear, suggesting a close relationship between these animals. The advantage to the wombat is that when a female is burrowing there is no danger of the pouch filling with earth and perhaps suffocating the baby. The advantage is less obvious for the koala for although the baby can easily get out to feed on the predigested pap flowing from the mother's anus, it has to grip tightly when returning to the pouch in order not to fall from the tree.

business. Already by 1924 the eastern Australian States were exporting more than two million koala pelts every year, entailing a rate of slaughter from which the population never really recovered. By 1927 the species had all but disappeared from New South Wales and Victoria; yet even this failed to prevent the authorities granting an increasing number of hunting licences annually. It was not until 1930, when it was abundantly clear that the koala was imminently threatened with extinction, that the first measures were taken to protect the animal. Fortunately the action came just in time and today the koala can be said to be out of danger, although it can never regain its former numbers, or anything approaching these.

## A gentle disposition

There are few animals anywhere that are more sociable and gentle by nature than the koala. All the naturalists who have had the opportunity of studying it at close quarters in the wild confirm that even when approached the animal maintains an air of peaceful nonchalance. Only if it is deliberately disturbed and acutely threatened will it display signs of alarm, scampering off to safety.

If taken at a sufficiently early age, the koala can be reared as easily as a pet cat or dog, although because of the specialised diet there is always the risk of it dying young. The ease with which the animal can be domesticated has in the past been a contributory factor towards a decline in numbers, but there are now strict laws in force to prevent people keeping it as a household pet.

The most celebrated case of adoption concerned a young female koala named Teddy who for some time lived and travelled about with Mr and Mrs Faulkner. Teddy's foster parents quickly realised that she tended to get miserable when, as inevitably happened, she had to be left alone. She would let out the most heart-rending appeals and could only be consoled by being picked up and cuddled in the arms like a baby. Eventually the Faulkners had the happy notion of making a cushion out of koala fur. This worked marvels and Teddy would immediately quieten down and happily sit clutching the object for hours. Later a toy teddy bear was substituted for the cushion and this worked equally successfully.

We have already seen in the chapter describing Dr Harlow's laboratory experiments with baby macaques the importance of a tactile relationship between a mother and her offspring, even if an artificial mother replaces the true one, and how the removal of this protective figure can lead to serious psychological disturbance. There seems to be an analogy here in the case history of Teddy the koala.

## The Australian possums

The Phalangeridae is one of the most diversified families of Australian mammals. Its representatives are the only Australian marsupials which have an almost entirely vegetational diet and

Although difficulty in obtaining the necessary food prevents most zoos exhibiting koalas, the peaceful little animals are among the world's favourites.

*Facing page:* Like other marsupials, the young koala is closely dependent upon its mother for some time after it leaves the pouch, straddling her back as she clambers up and down tree trunks. The advantage of this method of transport, as for many monkeys, is that her movements are not unduly hampered by the youngster.

The striped phalanger (*Dactylopsila picata*) is a marsupial yet it bears a striking resemblance to the aye-aye of Madagascar which is a Primate. Common features of the two species are a large upper incisor and the long fourth finger of the hand, used in both cases for removing grubs from tree bark.

*Facing page*: Among the members of the family Phalangeridae are various species of Australian possum (*above left*) and cuscus (*above right*). In the lower picture a baby possum suckles outside the pouch, clinging with its claws to the mother's fur so as to avoid falling.

which live in trees, occupying an ecological niche more or less equivalent to that of the tree-dwelling Primates of other continents. Apart from the tree kangaroo, which is one of the Macropodidae, the upper levels of the Australian forests are incontestably the exclusive haunts of the Phalangeridae.

Perhaps the most unusual members of this large family are the Australian possums, small animals with nocturnal habits which are not really named very appropriately, giving rise to some confusion with the American opossums. The Australians solve the problem by calling their animals possums.

The brush-tailed possum (*Trichosurus vulpecula*), also known in the fur trade as the Adelaide chinchilla, has a fox-like head and is enormously adaptable to a variety of habitats. Thanks to this the animal has been able to withstand intensive persecution (its beautiful fur, ranging from silver-grey to brown or black, being in great commercial demand) and the species is today present in large numbers, especially in suburban areas near bush.

As indicated, these Australian possums have a wide distribution. Trees are naturally their favourite refuges but they are also found in scrub, rock clefts and even under house roofs. They spend much of the day asleep and only venture out at nightfall.

Another species, the mountain possum (*Trichosurus caninus*), has darker fur and smaller ears than its relative, and lives at a higher altitude. Although fairly rare, this animal is not in such great demand for its pelt.

The largest and heaviest of these possums are the cuscuses, including the grey cuscus (*Phalanger orientalis*), the spotted cuscus (*Phalanger maculatus*) and the black cuscus (*Phalanger ursinus*). Because of their slow movements they recall the sloths of the Neotropical region; but the shape of the snout, the large protruding eyes and climbing habits remind one of monkeys.

Convergent evolution due to adaptation is most clearly in evidence in the case of the striped possum (*Dactylopsila trivirgata*), so named because of the distinctive black stripe that adorns the back and the flanks. This animal is astonishingly similar to the aye-aye of Madagascar, not so much in appearance as in the possession of a large upper incisor and a fourth finger which is longer and thinner than the others and is used for extracting insects from tree bark.

Referring to the honey possum or honey mouse (*Tarsipes spencerae*), W. Gewalt uses the description 'marsupial hummingbird', an apt comparison not only because this mouse-sized creature is the tiniest member of its family but even more because of an amazing similarity in feeding habits.

The honey possum (only about 3 inches long) has a slender snout, tiny eyes and a long, naked, prehensile tail. The strangest feature, however, is the bristly protractile tongue which can be extended about an inch beyond the muzzle and is used for sipping the nectar of flowers. This is the principal ingredient of its diet although insects also become attached to the tongue and are swallowed. It is an inhabitant of the sand-plain scrub of southwestern Australia, and the increasing agricultural use of this land threatens it in all parts of its range not within reserves.

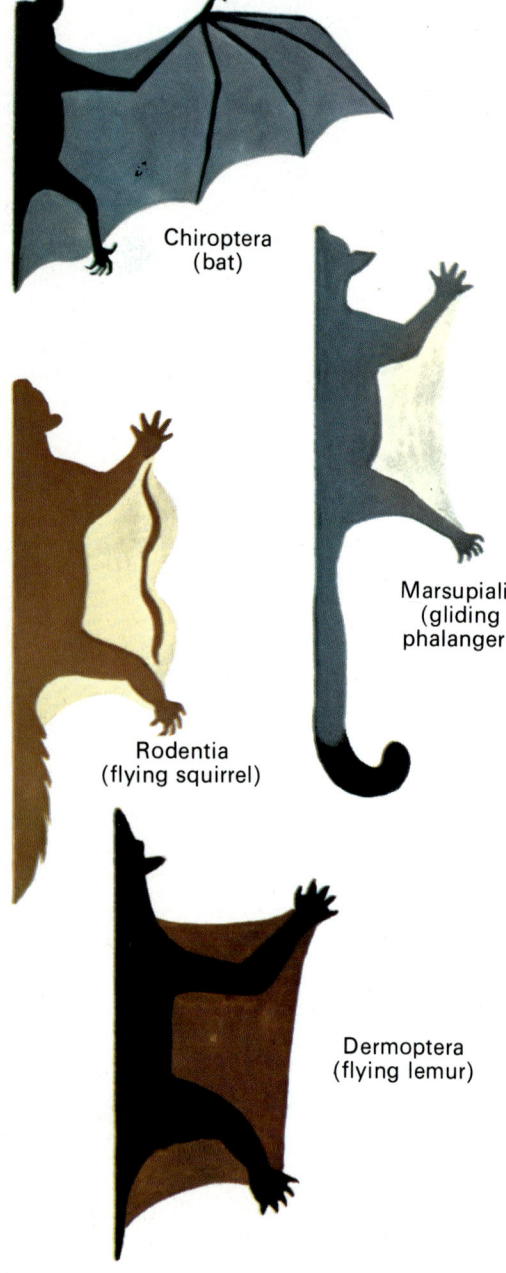

All flying and gliding mammals possess a skin membrane which acts as a kind of parachute. In the case of bats (Chiroptera), most accomplished of such fliers, it takes the form of a proper wing, supported by elongated fingers, leaving the hind toes free. In the Dermoptera, Rodentia and Marsupialia, animals which glide rather than fly – the extensible patagium varies in structure, usually linking the fore and hind feet and sometimes extending to the tail.

# Gliders of the forest

The Phalangeridae also include the gliding phalangers which, like the flying squirrels of other continents, possess a lateral skin membrane – the patagium – enabling them to glide from tree to tree.

In the case of the gliders this hairy membrane is situated along the flanks, linking the fore and hind feet. When the animal is at rest the patagium is folded but as soon as it extends its legs on either side the membrane automatically stretches open like an umbrella. The naturalist Ludwig Heck describes this as a remarkable sight, as if the animal's body had lost its third dimension and was suddenly transformed into a cloth or serviette.

The commonest of these parachuting marsupials is the short-headed or sugar glider (*Petaurus breviceps*), in size and appearance very like a flying squirrel. The animal has soft grey fur with a black stripe extending down the back from head to tail. Other species include the squirrel glider (*Petaurus norfolcensis*) and the yellow-bellied glider (*Petaurus australis*) which is about the same size as a domestic cat. Both species are rarely seen and difficult to study.

The pygmy or feather-tail glider (*Acrobates pygmaeus*) measures about 4 inches and it too possesses a delicate patagium for gliding from one branch to another. Despite the introduction of the domestic cat which is much given to hunting it, the numbers of this species do not seem to have declined to any appreciable extent, perhaps as a result of their tendency to conceal themselves whenever possible, and also their nocturnal habits. There are feather-tail gliders about in many bushland residential gardens but they generally go unseen, except when they dart out from their nest in a hole of a tree and make for another place of refuge. It is possible to be surrounded by these little ghost-like animals without ever suspecting their presence in the trees.

At the other end of the scale is the greater glider (*Schoinobates volans*). Measuring more than 3 feet, with a long bushy tail, this marsupial has a hugely extensible patagium, so that the animal is reputed to be capable of single glides of up to 80 yards. The structure of this flying membrane differs from that of related species in that it stems from the bend of the forelimb and extends down to the tip of the hind foot. Thus the silhouette of the glider in flight is slightly more triangular than that of the other gliders.

The greater glider is found only in the forests of eastern Australia, perched alone or in pairs at the summit of a gum tree. Like its distant relative, the koala, it is a selective feeder, concentrating on just a few species, notably *Eucalyptus australiana* and *Eucalyptus elaephora* in Victoria, and consuming leaves, shoots and flowers.

Survival has not been as difficult for the greater glider as it has been for other Australian animals. Its only regular predator is the powerful owl (*Ninox strenua*) but the activities of the latter appear to have relatively little impact on total numbers because for much of the time the gliders remain safely in the closed canopy at the top of a tree or asleep in nests in hollow branches. Probably a more important reason for their survival is that man is not interested in hunting the species. Although the long grey to dark brown fur is attractive enough, from the commercial point of view it is difficult to work and consequently not in demand.

## The skilful woodland burrowers

Of all the Australian marsupials only two, the wombat (family Vombatidae) and the marsupial mole (family Notoryctidae) are burrowers, leading a partially subterranean existence and occupying an ecological niche equivalent to that of the moles, viscachas and marmots of other continents.

The one species of marsupial mole (*Notoryctes typhlops*) differs in habit from true moles in that it does not live permanently underground in tunnels. Marsupial moles are essentially inhabitants of dry and desert regions whereas the wombats are found on the plains, and in both central and mountain forests. Marsupial moles lack both eyes and external ears. Their fur is silky, ranging from white to russet, and two of the claws on the forefeet are elongated for burrowing. Very little is known of their behaviour. The wombats, being much larger, are less secretive and have consequently proved much more rewarding from the scientific point of view.

The wombat was first discovered by the white man towards the end of the 18th century by sailors whose ship had capsized in the Bass Strait, between Australia and Tasmania, and who had taken refuge on an island. The place was apparently alive with

The large patagium of the greater glider (*Schoinobates volans*) makes it possible for this bushy-tailed phalanger to glide up to 80 yards at a time. The size of a cat, this inhabitant of eastern Australia (also shown in the lower picture) moves clumsily on the ground.

these animals and they managed to survive by eating them. When the rescue party arrived they took one along to show to their companions, describing them, not very imaginatively, as 'a species of wild animal'.

Some years later the white colonists rediscovered the wombat – this time on the mainland. Although there is some disagreement regarding classification so that some wombats are variously listed as species or subspecies, scientists have divided them broadly into two groups – naked-nosed and hairy-nosed wombats. The only naked-nosed species is the common wombat (*Vombatus hirsutus*) of the eastern forests, and is characterised by coarse dark brown hair (the colour is variable) short ears and a bare patch on the muzzle.

Rather smaller, with greyish-brown fur, large ears and a hairy muzzle is the best known representative of the second group, the hairy-nosed wombat (*Lasiorhinus latifrons*).

The range of distribution of naked-nosed wombats is south-eastern Australia, Tasmania and Flinders Island (between Tasmania and the mainland). The hairy-nosed wombats are to be found on the flat savannah grasslands of South Australia and southern Queensland.

These marsupials, which look a little like marmots or badgers, dig large burrows with the strong claws of their forefeet. The tunnels may extend deep below the ground but usually run for considerable distances more or less parallel to the surface. Inside are the nests which are lined with bark, grass and leaves.

Wombats live on their own and the sexes only come together in the breeding season, between April and June. The nests of a pair of wombats do not communicate with those of neighbours. Sometimes small depressions are visible in the ground near the entrance of a burrow. These are used for sleeping and for sunbathing during the day for the animal ventures out for food only at night, shuffling along well-worn paths on its stocky legs rather like a bear. The food of wombats consists in the main of grass, bark and tree roots.

The burrow is important in that it provides protection against roaming ground predators. Should any enemy try to enter the tunnel the wombat performs an abrupt about-turn, presenting its rump to the intruder. Since the skin is usually thick and tough and the animal has no tail the predator has no way of inflicting an injury. Alternatively the wombat will block up the entrance to the burrow by hunching its back and this too prevents the enemy striking a blow at a really vulnerable portion of the body. Should this ruse fail the marsupial can kick out with its little feet and try to scratch the predator with its claws.

The wombats have been accorded separate status in the family Vombatidae. The reason for their being classified apart from other marsupials is that the structure of their teeth more resembles that of rodents, with two incisors to each jaw. In other respects they are clearly akin to koalas and some experts consider that the two groups are closely related. For example, they possess a long caecum which, although smaller than that of the koala, evidently plays an important part in the digestion of the miscellaneous vegetation which they eat. Furthermore, the

The large squirrel glider (*Petaurus norfolcensis*) is now a rare species and for that reason its habits have been little studied.

*Facing page:* The short-headed gliding phalanger, otherwise known as the sugar glider (*Petaurus breviceps*), is a nocturnal animal which, like other members of its family, feeds on vegetation but also eats insects. It has a wide range of distribution, including Tasmania, and is probably found in greater numbers than any other gliding marsupial.

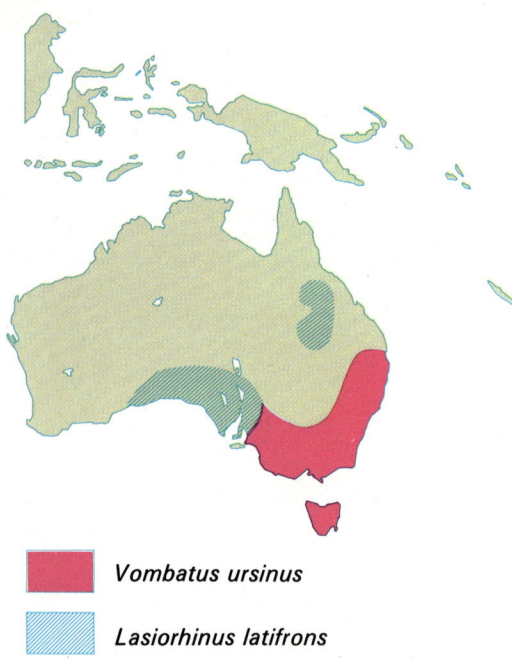

Geographical distribution of the common wombat (*Vombatus ursinus*) and the hairy-nosed wombat (*Lasiorhinus latifrons*).

---

**WOMBATS**
Class: Mammalia
Order: Marsupialia
Family: Vombatidae

**COMMON WOMBAT**
(*Vombatus ursinus*)

Length of head and body: 27½–47½ inches (70–120 cm)
Tail: in embryo only
Weight: 33–77 lb (15–35 kg)
Diet: leaves, grass, roots, bark, etc.
Longevity: 25–30 years

Belongs to naked-nosed group of wombats. Coarse fur, varying in colour—yellowish, brown, even black. Stocky body, large head, slightly pointed ears. Two teats in marsupium which opens towards rear.

**HAIRY-NOSED WOMBAT**
(*Vombatus* or *Lasiorhinus latifrons*)

Length of head and body: 34½–40 inches (87–102 cm)
Length of tail: 2 inches (5 cm)
Weight: 55–66 lb (25–30 kg)

Similar to naked-nosed wombats but fur is softer and usually darker in colour, ears larger and nose hairy.

---

*Facing page:* The wombats are the most specialised burrowing animals in Australia, digging long tunnels for their nests. Zoologists divide them into two groups, the naked-nosed wombats (*below*) and the hairy-nosed wombats (*above*).

---

marsupium, as with the koala, opens towards the rear, another feature which suggests to some authors that the two animals may have had a common ancestor.

The functional role played by the rear-opening pouch is more obvious in the case of the wombat. For one thing this animal normally proceeds along the ground on all fours so that the position of the body is different from that of the upright-perching koala. In addition, when a female is burrowing in a tunnel the fact that the marsupium inclines backwards avoids the possibility of it becoming filled with soil and perhaps injuring the baby that she is carrying. This would probably happen if the pouch were open towards the front.

Wombats are extremely timid, cautious animals yet they adjust well to life in captivity. In their case the feeding problems are not as great as they are for koalas. They too are capable of responding to human companionship and can be reared as domestic pets.

Like most Australian mammals, wombats have been much persecuted. In the early days, prior to the introduction of imported species, they lived a relatively peaceful existence, troubled only by the Aborigines who killed them for food. But when large tracts of land were given over to farming the little animals were hunted mercilessly, on the pretext of damage allegedly caused to fencing and grazing grounds. It was also claimed that livestock risked breaking their legs by stumbling accidentally into wombat burrows and this provided an even more outrageous excuse for prosecuting what amounted to all-out warfare against the peaceful and totally inoffensive animals.

The situation took a marked turn for the worse when imports of rabbits occurred. These multiplied at such a rate that they were soon nibbling pastures bare, depriving the wombats of their food and even penetrating their subterranean tunnels. The obvious method of combating the disproportionate growth of the rabbit population was to lay poison but as often as not this killed as many wombats as rabbits. Despite these pressures, wombats still remain common in many areas of south-east Australia, particularly in the forested hills of the Great Dividing Range.

As a species the common wombat is probably secure because in part of its range it occupies high mountain forest, where it is left in peace. The hairy-nosed wombat, by contrast, seems to be occupying a rapidly shrinking range. It is locally common but vulnerable because it lives in densely populated communities. The habitat of the hairy-nosed wombat is woodland savannah or grassy plains of the dry interior: on the plains to the west of Murray River, the southern border of the Nullabor Plain and the Gawler Range of South Australia. One colony, at Blanchetown, in South Australia, is protected in a reserve of 3,000 acres.

R. T. Wells, of the University of Adelaide, has studied the hairy-nosed wombat in captivity. He finds it has the characteristics better known in small desert animals elsewhere, such as the kangaroo-rat of the south-western United States and the gerbils and jerboas of Asia and Africa, of being able to go long periods,

Common wombat
(*Vombatus ursinus*)

Hairy-nosed wombat
(*Lasiorhinus latifrons*)

*Facing page*: Although the echidna or spiny anteater superficially resembles a hedgehog, it is one of the most primitive of living mammals, together with the duckbill or platypus. Both are egg-laying mammals belonging to the order Monotremata.

even perhaps indefinitely, without drinking, and for similar reasons. The wombat stays in its burrow avoiding the heat of the day, emerging at sunset and dawn to feed. This enables it to avoid loss of body water by evaporation from the skin. It does not sweat and its droppings are dry, so the loss of water is minimal. The urine is highly concentrated and semi-solid.

Another adaptation of the hairy-nosed wombat to desert conditions is that this animal can subsist on a diet low in protein, such as dry plant material. Its stomach contains micro-organisms capable of digesting cellulose and making it available as food to the wombat.

## The primitive echidna

A layman travelling through the Australian bush may be forgiven for not paying too much attention to the echidna or spiny anteater, for this small animal can easily be confused with a common hedgehog. But in fact the echidna deserves to be treated with great respect and consideration for it is one of the two most primitive mammals in the world, the other being the duckbill or platypus.

Although similar in outward appearance, echidnas belong to two genera. Those of the genus *Tachyglossus* possess a long, straight muzzle, the two constituent species being the Australian echidna (*Tachyglossus aculeatus*) and the Tasmanian echidna (*Tachyglossus setosus*). The three representatives of the genus *Zaglossus*, on the other hand, have a curved muzzle. They are Bubu's echidna (*Zaglossus bubuensis*), Barton's echidna (*Zaglossus bartoni*) and Bruijn's echidna (*Zaglossus bruijni*).

The echidnas of the former genus are smaller and have shorter tails than the latter, but the three species of *Zaglossus* are faster moving.

All echidnas are to be found in districts with good tree and plant cover, especially on rocky soil. They venture out around dusk and then spend the entire night rummaging for food, this consisting of termites and ants which they trap with their long, sticky tongue as do the anteaters of Africa and Asia.

Since their vision is rather poor, echidnas track down their prey by smell, probing indefatigably at the ground with their beak-like muzzle. Like the true anteaters their feet are furnished with strong claws for digging (five on each foot). These are wonderfully effective for burrowing or for ripping apart a termite mound.

The sharp spines that envelop the upper part of the echidna's rounded body are its principal weapons against predators. By contrast, the abdomen is hairy and thus much more vulnerable. Because of this anatomical peculiarity the spiny anteater, either when threatened or simply when it settles down to sleep, rolls itself into a tight spiny ball, or digs a hole with its claws at an astonishing speed and partially buries itself, leaving only the spiky top of the back exposed and gripping the ground so firmly that it is almost impossible to dislodge it. If in imminent danger it will complete its burrowing activities in a matter of seconds and disappear completely.

The echidna or spiny anteater feeds on ants and termites which it catches with the long sticky tongue shooting out from its beak-like muzzle.

Undoubtedly the most interesting characteristic of the echidna, which most clearly reveals the animal's primitive nature, is its strange method of reproduction. In contrast to other mammals, the Monotremata do not possess separate orifices for the uro-genital and digestive functions so that urine, feces and sexual secretions are all expelled through a single tract, the cloaca. Furthermore the females do not have a vagina or even a rudimentary uterus so that it is impossible for the fetus to develop inside the mother's body. Since the impregnated female is unable to provide the necessary warmth and protection herself, nature's solution to the problem has been to continue the reproductive methods of reptiles and birds. The one-celled embryo is surrounded by a tough protective membrane or shell; in other words the baby hatches from an egg.

In the case of the echidna the egg is ejected from the cloaca and then passes directly into the pouch for incubation. Unlike the pouch of marsupials this appears on the female's abdomen only after she conceives. She possesses no teats but inside the pouch are diffuse mammary glands, no more than slits, out of which the maternal milk oozes.

Experiments on anaesthetised female echidnas have shown that slackening of the abdominal muscles causes the incubation pouch to disappear, indicating that the latter is just a fold in the skin of the belly. Exactly how it is formed remains a mystery. So too does the manner in which the egg finds its way into the pouch, for there is absolutely no evidence that the echidna makes use of her feet, nor does she bend her body in such a way that the pouch is placed opposite the cloaca. Some authors are of the opinion that she rolls the egg into the pouch with her beak; the most recent view is that the egg is placed in the pouch by a protusible cloaca.

The actual hatching process, whereby the baby breaks out of its imprisoning shell, occurs inside the pouch. Like certain birds, the echidna has a hard projection, known as the diamond, at the tip of its beak and this is used by the baby for cracking the shell open.

After eight to ten weeks the baby echidna's spines begin to grow. The contact of the sharp spines against the abdomen makes life uncomfortable and eventually insupportable for the mother, who turns the baby out of the pouch but deposits it in a safe place of refuge, visiting it at regular intervals for suckling. A year after leaving the pouch the echidna is fully adult.

The survival of such a primitive animal as the echidna in Australia is due to several factors. In the first place it has remarkably effective ways of defending itself against enemies—sharp spines to inflict injury and an ability to bury itself underground with disconcerting speed. Secondly, the animals imported by the white settlers did not in any way threaten their food supply. The final reason is that the echidna may live for fifty years or more so that one female will have numerous offspring. Consequently the curious echidna is not at the moment in danger of becoming extinct.

## The tree-kangaroos

It must have come as quite a shock to the zoologists who first had the opportunity of studying the forest fauna of northern Queensland and New Guinea to be confronted suddenly with kangaroos perching in trees. After all, this marsupial, as was well known, was essentially an animal of the wide open spaces, with long powerful legs ideally suited to its characteristic leaping style of movement. Yet here was a kangaroo which lived in the heart of the forest and which was evidently a climber rather than a jumper.

The seven species of tree-kangaroo belonging to the genus *Dendrolagus* differ surprisingly little in outward appearance from their relatives of plain and savannah. Their ears, however, are smaller, there is less difference in the dimensions of the fore and hind feet (all four being provided with broad soles and strong, curved claws), and the fur is rather longer and thicker. The tail too is thinner and less muscular than that of the ordinary terrestrial kangaroo.

These minor anatomical variations cannot really be described as adaptations to an arboreal way of life. The fur merely provides some extra protection against rain, which may be heavy in these forest regions. The modest size of the tail indicates that it is not employed as a third point of support for the body, as is the case with the majority of kangaroos, and certainly it is not prehensile. Only the feet, with their broad sole surface and well developed claws, would appear to be of genuine value to an animal which spends some, though not all of its time, in trees.

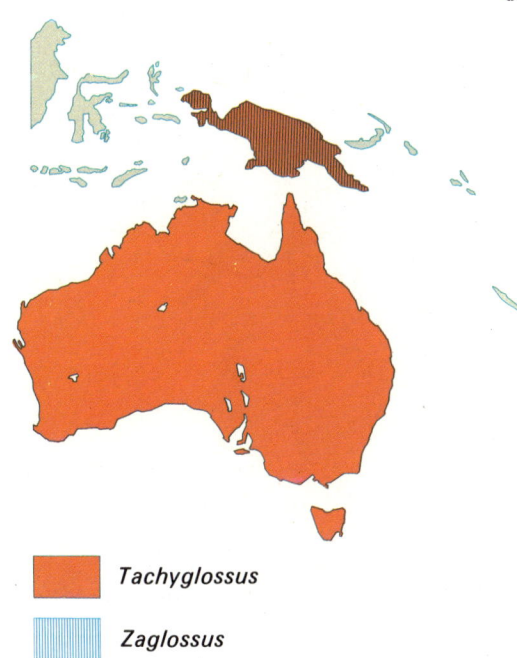

▮ *Tachyglossus*

▮ *Zaglossus*

Geographical distribution of straight-muzzled echidnas (*Tachyglossus*) and curved-muzzled echidnas (*Zaglossus*).

---

**ECHIDNAS**

Class: Mammalia
Order: Monotremata
Family: Tachyglossidae

**AUSTRALIAN ECHIDNA**
(*Tachyglossus aculeatus*)

Length of head and body: 14–21 inches (35–53 cm)
Length of tail: 3½ inches (9 cm)
Weight: 5½–13 lb (2·5–6 kg)
Diet: termites and ants
Number of young: one
Longevity: 50 years or more

Body, with brown or blackish pelage. Back covered with 2-inch pointed spines. Long, almost straight, beak-like muzzle; protrusible tongue. Five toes on feet, all with strong claws. Related species is Tasmanian echidna (*Tachyglossus setosus*).

**BRUIJN'S ECHIDNA**
(*Zaglossus bruijni*)

Length of head and body: 18–31½ inches (45–80 cm)
Length of tail: 3¼–3½ inches (8–9 cm)
Weight: 11–22 lb (5–10 kg)
Diet: termites and other insects
Number of young: one
Longevity: 30–50 years

Lighter in colour than Australian species, with shorter spines and a longer, more curved muzzle. Other species, also from New Guinea, are *Zaglossus bartoni* and *Zaglossus bubuensis*.

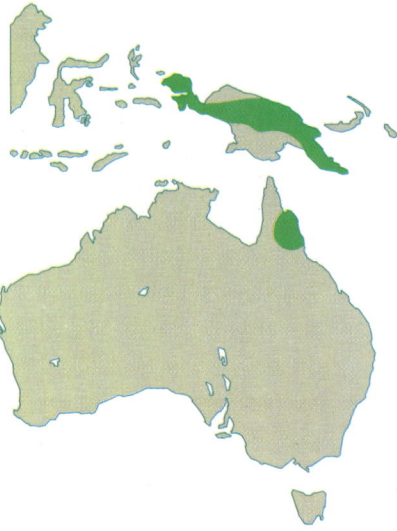

Geographical distribution of tree-kangaroos.

---

**BLACK TREE-KANGAROO**
(*Dendrolagus ursinus*)

Class: Mammalia
Order: Marsupialia
Family: Macropodidae
Length of head and body: 20½–32 inches (52–81 cm)
Length of tail: 16½–37 inches (42–94 cm)

Long, thick fur, dark brown above, becoming paler on abdomen and white on throat and beneath jaws. Forelimbs better developed and hind limbs shorter and less strong than those of terrestrial kangaroos. Strong hooked claws. Other tree-kangaroos are *Dendrolagus lumholtzi*, *Dendrolagus goodfellowi*, *Dendrolagus matschiei* and *Dendrolagus bennettianus*.

---

The climbing ability of the tree-kangaroo means that it has a wide range of vegetation to feed upon, in a habitat where it need fear no large predators.

Strong soles and sharp claws enable the tree-kangaroos to move about freely above ground level. Otherwise only their smaller ears, more slender tail and denser fur distinguish them outwardly from terrestrial kangaroos. The picture shows a female black tree-kangaroo and her joey.

Lumholtz's tree-kangaroo (*Dendrolagus lumholtzi*)

Bennett's tree-kangaroo (*Dendrolagus bennettianus*)

Matschie's tree-kangaroo (*Dendrolagus matschiei*)

The truth is that the tree-kangaroo is not a particularly agile animal, a point noted by Alfred Russel Wallace when he described the genus. He commented that the tree-kangaroo differed very little from its terrestrial relative and that it appeared to be poorly adapted for climbing and not well suited to an arboreal existence, moving slowly and ponderously, and clinging very precariously to branches.

Since survival depends to a large extent on adapting to specific conditions and habitats, how is it that the tree-kangaroo has managed to surmount such problems without apparently possessing the physical means to do so? How has an animal, specialised in a leaping style of locomotion, transformed itself into a climber without undergoing any significant change of appearance? The answers are quite simple.

The forests of the Australasian region offer a much greater quantity and variety of food than can be found on the plains and savannahs, particularly for an animal as large as a kangaroo which enjoys the double advantage of having no serious rivals or dangerous enemies. The tree-kangaroo's feet and claws are sufficiently strong for clambering up a tree in search of food and its legs powerful enough to make long leaps to safety if suddenly alarmed. The selective pressures which normally give rise to the physical or physiological mechanisms necessary for survival are here absent. So the tree-kangaroo, which spends much of its time on the ground, has adapted to its particular surroundings rather than to a general mode of life.

# CHAPTER 99

# Birds of the Australian woodland and savannah

Travellers to distant countries are notoriously prone to exaggeration. Even when there is no intention to deceive the listener or reader, the free embroidery of the imagination often makes it hard, if not impossible, to distinguish legend from fact. The pattern, of course, was set in days long past when exploration was in its infancy and superstition rife. Conceivably unicorns, sirens and phoenixes really could have existed. Even today there are those who have not given up the search for the abominable snowman.

Thus it was with the apparently legendary birds of paradise described by Spanish sailors who had sailed with Magellan. True, they produced some evidence—stuffed birds, lacking bones or feet, but with glowing feathers—presents, they swore, from the little dark skinned men of New Guinea. According to myths circulating in the Spice Islands these fabulous birds were lords of the air, never landing on the ground or perching in the branches, simply hovering to feed on nectar and flower petals. It was said that they nested in Paradise itself.

Could such unlikely stories be believed? Three centuries later navigators exploring the Pacific discovered to their delight that these birds really did exist. Seeing them in their natural habitat, naturalists reported that the unimaginably beautiful feathers well merited them the name of birds of paradise, but it was obvious that in other respects the ancient legends and earlier descriptions could not be true. For although the birds were undeniably handsome, their bearing proud and graceful and their courtship display breathtakingly spectacular, they could not by any stretch of the imagination be considered rapid fliers. Aristocrats of the forest they might be; lords of the sky they were not.

*Facing page:* The woodland regions of Australia are rich in birdlife. One of the strangest species is the flightless cassowary, slightly smaller than the emu, recognisable by its black plumage, bare red and blue patches on head and neck, wattles hanging from the throat, and bony casque on the forehead.

There were, however, many other surprises in store. Sailors who had set foot on the Australian mainland told of birds that were capable of astonishing constructional feats, building conical bowers, arbours, avenues and gardens out of bits of vegetation and actually decorating them with flower petals, seeds, snail shells, tiny bones and glittering stones. Even more incredibly, they would 'paint' the interior of their chamber by crushing certain plants to pulp, mixing the extracted coloured juice with saliva and applying it with a 'brush' of vegetation held in the beak.

Such reports were no mere figments of the imagination for sceptical ornithologists soon verified with their own eyes that the satin bowerbird, perhaps the most remarkable member of the group, does in fact manipulate a wad of bark in precisely this way to decorate the inner walls of its bower.

Naturalists very quickly realised that the birdlife of Australasia was uncommonly rewarding. In New Guinea they came across a pigeon as large as a peacock, with a superb head crest. They

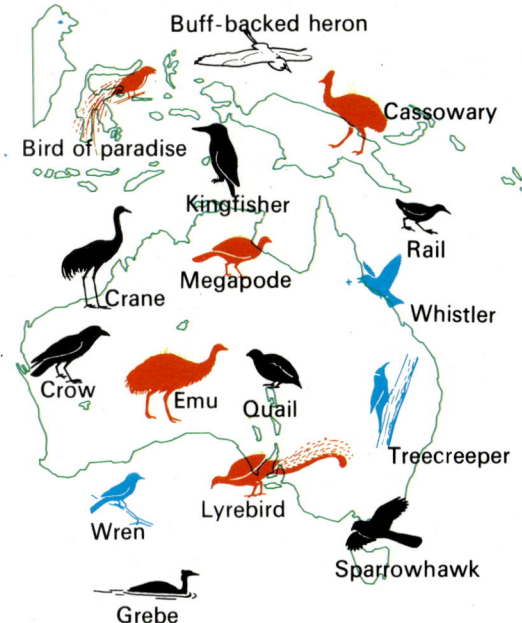

The birds of the Australasian region reached their present island homes at different times, some of them being able to cross broad stretches of water, others making use of land links that no longer exist. Among the first arrivals were birds of paradise, cassowaries, emus, megapodes and lyrebirds – all of them endemic groups – and marked here in red. Later groups included rails, cranes, quails, crows, sparrowhawks, grebes and kingfishers (shown in black), followed by wrens, whistlers and treecreepers (in blue). The most recent arrival has been the buff-backed heron or cattle egret (in white).

*Left and facing page:* The kookaburra or laughing jackass, whose distinctive nest in a termite nest is illustrated here, is a member of the kingfisher family and is noted for its curious cries which sound like peals of human laughter. Its food includes rodents and reptiles.

*Right and facing page:* Among the most widely distributed groups of Australian Passeriformes are the flycatchers of the family Muscicapidae, notable for their handsomely coloured plumage. The birds illustrated here are the crimson chat, *Epthianuta tricolor*, (*top centre*), the variegated wren, a warbler, *Malurus lamberti*, (*right*), the grey fantail, *Rhipidura fuliginosa*, (*bottom left*) and the yellow flycatcher, *Microeca flavigaster*, (*top left*). Other Passeriformes include the honeyeaters, Meliphagidae, and bowerbirds, represented here respectively by the white-naped honeyeater, *Melithreptus lunatus*, (*bottom centre*) and the green catbird, *Ailuroedus crassirostris*, (*centre left*).

The honeyeaters of the Australian forests feed almost exclusively on nectar, sipping it with the tongue which curls up at tip and sides to form a tube.

examined with fascination the curious incubation mounds of the megapodes. They marvelled at a large bird whose magnificent plumage was matched by its varied and melodious vocal repertory, which included mimicry of other species – the lyrebird; and they were amazed to find that parrots, parrakeets and cockatoos, were here present in enormous multicoloured flocks.

Although South America has the reputation of being the continent that is numerically the richest in birdlife, Australasia can certainly boast the most unusual species. Geographical isolation, the absence of predators and a pleasant climate have combined to produce a host of evolutionary marvels and curiosities. Even without delving into the scientific explanations, the gorgeous hues of the birds of paradise, the decorated bowers and avenues of the bowerbirds, the artificial incubators of the megapodes, the crest of the blue-crowned pigeon and the vocal dexterity of the lyrebird are as wondrous as anything dreamed up in the fabulous world of Sinbad the Sailor.

## The avifauna of Australasia

There is a broad similarity between the birds of the different parts of the Australasian region, so that although some zoogeographers regard Australia and New Zealand as separate regions, in this respect they are better treated as subregions. In proportion to its area Australia in fact has the fewest bird species. There are just as many (about 750) in the much smaller island of New Guinea, and slightly over 1600 in the two islands combined. New Zealand has about 215 species.

Australia was separated from Asia at the beginning of the Tertiary epoch or even a little before that, forming a solid land mass with New Guinea. It is not known when Australia was severed from New Zealand. The avifauna of Australasia undoubtedly has close affinities with that of South-east Asia but long isolation has given rise to much differentiation among endemic groups and to the formation of numerous families. Only the Neotropical region possesses more endemic species than Australasia. It is worth mentioning that although there are a number of groups that have come to Australasia from the Palearctic region there is no such interchange with the Neotropical region.

The majority of Old World groups are represented in one form or another in Australasia, but some are missing, including, among others, vultures, flamingos, sand grouse, scissorbills, pheasants (quails excepted), woodpeckers, finches and buntings. Some of the more ambitious migratory species do touch Australia in the course of their amazing journeys (notably two species of swifts, a cuckoo and a swallow) but the majority of Passerines do not usually get so far.

There is a high proportion of endemic species—birds that are not found anywhere else—in Australia. The longer they have been here the more diverse the species, but in fact it is difficult to decide how long they have lived in isolation. Among the most characteristic are cassowaries and emus, megapodes, cockatoos, kiwis, lyrebirds, butcherbirds, mudlarks (or magpie-larks), wood swallows, honeyeaters and bowerbirds. More recent arrivals have given rise to equally distinctive groups but with obvious resemblances to those of the Old World. They include flycatchers, wrens and treecreepers. Other groups of birds are clearly recent immigrants and are common elsewhere, such as quails, rails, divers, cranes, sparrowhawks, crows, etc. New species continue to arrive, the last being the buff-backed heron.

Some of the most widely distributed species have attained great popularity in Australia. One well known example is the kookaburra or laughing jackass (*Dacelo gigas*), a member of the kingfisher family whose name is derived from its strange cries, uncannily like bursts of human laughter. In common with other kingfishers the kookaburra is a tree-dwelling hunter which waits in ambush on a branch for its prey. Large insects are caught in mid-air while small vertebrates, mammals and reptiles, are taken on the ground, the kookaburra diving down like a lanner falcon. Crabs are also caught in this manner. Since it is incapable of tearing up large prey prior to swallowing them, the bird batters them into a pulp with its large bill. Because of its food preferences (especially its snake-killing propensities) the laughing jackass is tolerated by farmers and has in fact been introduced to certain islands, including Tasmania.

The 160 species of honeyeaters (Meliphagidae), which are

The tribesmen of New Guinea have for centuries worn the gorgeous feathers of birds of paradise as facial ornaments and head-dresses.

Lawe's bird of paradise
(*Parotia lawesii*)

Magnificent bird of paradise (female)
(*Diphyllodes magnificus*)

Magnificent bird of paradise (male)
(*Diphyllodes magnificus*)

Passeriformes, are also extremely widely distributed. Most of them are rather drab in colour but some have bold black, white and red patterns. The most characteristic feature of the honeyeaters is their choice of food and method of obtaining it. The bill is slender, long and curved, and the extensible tongue, covered with bristles, curls up at the sides and tip to form a kind of tube. It is through this tube that the honeyeater sucks nectar from flowers as well as swallowing insects that happen to be feeding on the flowers at the time.

The constant attentive visits of the honeyeaters to various plants and the contact between the flower organs and the birds' feathers play an important part in pollination and propagation, especially of *Eucalyptus* species. Botanists have discovered a series of interesting adaptations in some of these trees, which appear to have no other purpose than to facilitate pollination by such means.

## The amazing courtship dances of birds of paradise and bowerbirds

The most unusual and awe-inspiring courtship rituals among the birds of Australia are those performed by the birds of paradise (Paradisaeidae) and the bowerbirds (Ptilonorhynchidae). Both entail displays of exceptional beauty and complexity, indicating a relatively high level of intelligence. Although not universally agreed, the birds of Australia seem to be more advanced from the evolutionary viewpoint than the mammals.

The ornithologist who has made the most profound study of the breeding behaviour of these two families is Thomas Gilliard of the American Museum of Natural History. As a result of his many journeys to the forests of New Guinea he has provided us with a comprehensive survey of the fascinating courtship habits of these birds, and the information contained in this section is largely based on his findings.

The birds of paradise, which are divided by Mayr into two subfamilies, are mainly concentrated in New Guinea and adjacent islands, whilst a few are to be found in the Moluccas and in north-eastern Australia. They comprise about 20 genera and 43 species. Although very closely related to crows, being heavily and strongly built with solid feet and thick bills, the outward resemblances end there, for the birds of paradise are mostly distinguished by the splendid colours of the plumage of the males and a wonderful diversity of ornamental feathers.

Essentially tree-dwellers, these magnificent birds are inhabitants of the rain forests and many species live only in mountain regions. Their diet is varied, some having a marked preference for fruit and confining their visitations to certain trees, some feeding on insects, and others concentrating on vertebrates, especially frogs and small arboreal lizards.

Solitary by habit, birds of paradise do not come together except during the breeding season. Then they assemble on a patch of ground variously known as an arena or stage, which is reserved exclusively for nuptial displays. Within this space each cock jealously guards his own little piece of territory, which he may have won by fighting and conquering a rival. The parcels of territory in the centre of the arena appear to be the most coveted.

As a general rule sexual dimorphism is very striking among birds of paradise, the drab, sober feathers of the females contrasting strongly with the vivid hues of the males. The plumage of the latter is not only richly coloured and often delicately patterned in glowing yellows, reds, emeralds, purples and the like, but accentuated by ornamental feathers that may assume the most extraordinary forms.

Among these adornments are delicate, filmy remiges that extend beyond the tail like long thin wires, flag-plumes that trail from the head (so called because of the miniature 'flags' that are attached to the vanes), and feathered bibs and gorgets or feathers that spread out over the back or shoulders like sumptuous silken mantles. In many species the small, tightly clustered head feathers take on the appearance of glossy scales. Contrasting light and dark shades also heighten the overall effect, yet some rare species are black with only a few brilliant touches of colour here and there, and bright fleshy wattles.

Although men have long admired the beauty of birds of

Contrasting colours, extraordinary ornamental feathers and an acrobatic courtship dance from the branches make the greater bird of paradise a truly spectacular species.

Red-plumed (Count Raggi's) bird of paradise (*Paradisaea raggiana*)

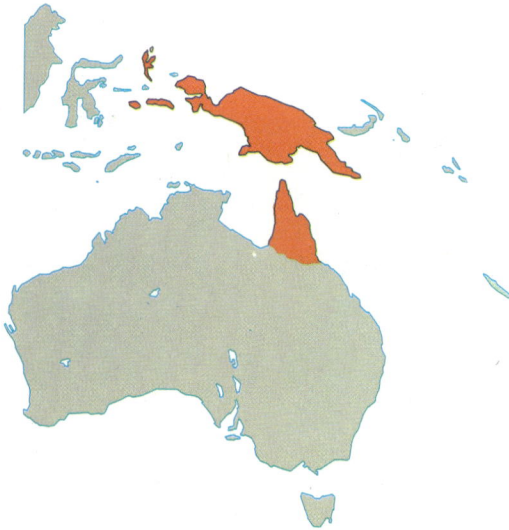

Geographical distribution of birds of paradise.

---

### BIRDS OF PARADISE

Class: Aves
Order: Passeriformes
Family: Paradisaeidae

#### PRINCE RUDOLPH'S BLUE BIRD OF PARADISE
(*Paradisaea rudolphi*)

Total length: 25 inches (63 cm)
Diet: fruit, arthropods and small vertebrates
Number of eggs: 1 or 2

Male has jet-black head, throat and neck with metallic reflections. Powerful white bill; white surrounds to eyes. Flanks adorned with long silky feathers, some blue, others cinnamon, which are very strong with separated barbs. Short tail has two large brown rectrices, terminating in blue tufts. Female is smaller and drabber.

#### EMPEROR OF GERMANY'S BIRD OF PARADISE
(*Paradisaea guilielmi*)

Total length: 32 inches (81 cm)
Diet: fruit, arthropods
Number of eggs: 1 or 2

Male has emerald green head and throat; bright yellow nape, shoulders and wing coverts which are small, forming two lateral bands over the breast. Brown wings, belly and rump. Medium-sized tail with two central wire-like rectrices. Yellow iris. Female is mainly brown with ochre shoulders and nape; front of head and neck are blackish.

paradise, it is only quite recently that reliable information has been forthcoming about their behaviour. This, in its way, is every bit as remarkable as their appearance. Studies in the wild of their courtship and breeding habits have now shown, for example, that there is a strong link between the exhibitions that take place during the breeding season and the phenomenon of sexual dimorphism. In those species where there is not a clearcut distinction in outward appearance between the male and the female there is a marked tendency on the part of the cock towards monogamy. The male helps the female to construct the nest and assists her in subsequent maternal duties. In the majority of species, however, the male will be much more brilliantly garbed than the female and will probably be polygamous. His interest in her lapses as soon as courtship and mating are done and he will not assume any paternal responsibilities.

Some species prepare a nuptial chamber on the forest floor, in the undergrowth or on the lower branches of a tree, and in such cases the courtship displays are private affairs. Other species flaunt themselves in the treetops and quite a number of individuals may engage simultaneously in such activities. But no matter where the nuptial dance occurs, the intention of each cock is the same, to show off his splendid colours and ornamental feathers to best advantage. Such feathers are erected and outspread, and the most richly adorned parts of the body exhibited, often to the accompaniment of strutting dance steps and strident cries. Some birds of paradise which display themselves on a branch will even hang by their feet head-downwards, whilst others adopt equally strange positions although maintaining an upright stance.

The majority of species build enormous bowl-shaped nests of branches and twigs, situated in a tree. One exception is the greater bird of paradise (*Paradisaea apoda*) which constructs its nest in a tree cavity. The normal number of eggs is one or two.

The marvellous feathers of birds of paradise have long been prized for their ornamental value and for other more practical purposes as well. The tribesmen of New Guinea and the Moluccas have not only used them for adorning clothing and as ceremonial head-dresses but also as a form of money to be exchanged in commercial transactions. The trade in plumage was never so extensive as to menace the future of any species, but when the Europeans began to take an interest the situation threatened to become serious. It was in the early 16th century that Spanish sailors first brought home samples of these birds' plumage and this later stimulated a commercial demand which swelled and receded in the centuries to come according to the whims of fashion. This fad brought some species under severe pressure. Today the conservationists have intervened to save the birds of paradise. Although they have not managed to persuade native tribes to renounce the habit of adorning their persons with these gaudy feathers, they have put a stop to the much better organised and potentially dangerous export trade in these commodities.

Ornithologists, having identified several dozen species of birds of paradise, were struck by the fact that a variety of

Prince Rudolph's blue bird of paradise
(*Paradisaea rudolphi*)

Lesser bird of paradise
(*Paradisaea minor*)

Arfak six-wired bird of paradise
(*Parotia sefilata*)

Twelve-wired bird of paradise
(*Seleucidis melanoleuca*)

*Facing page:* Ornamental features of the males of certain birds of paradise include long wire-like tail feathers. The species illustrated here are the magnificent bird of paradise (*above*) and Prince Rudolph's blue bird of paradise (*below*).

Geographical distribution of bowerbirds.

---

**BOWERBIRDS**

Class: Aves
Order: Passeriformes
Family: Ptilonorhynchidae

**STRIPED GARDENER BOWERBIRD**
(*Amblyornis subalaris*)

Total length: 9½ inches (24 cm)
Diet: fruit, arthropods, molluscs
Number of eggs: 1–3

Chestnut plumage; large orange-red crest which lies flat along back when not erect; brown eyes.

**SPOTTED BOWERBIRD**
(*Chlamydera maculata*)

Total length: 11–11½ inches (28–29 cm)
Diet: fruit, arthropods, molluscs
Number of eggs: 1–3

Light brown plumage with black flecks. Male has small pink crest on nape.

---

intermediate forms existed and these were eventually found to be wild hybrids—offspring of two interbreeding species. This tendency led to a considerable amount of confusion among ornithologists at the beginning of this century, with hybrids classified as separate species. Since the males involved would have been promiscuous, belonging to species with strong sexual dimorphism, differences in colour and ornamentation between male and female are usually evident in these hybrid forms.

## The astonishing bower and avenue builders

The nineteen species of Australasian bowerbirds are related to the birds of paradise. Six are exclusively Australian, two are found both in Australia and New Guinea, and eleven are inhabitants of New Guinea alone. Some are as small as thrushes, others as large as crows. Their wings are rounded and the colour of their plumage is very variable.

The bowerbirds are unique among Passeriformes in their amazing ability to distinguish and utilise colours, their architectural skills and their use of objects for decoration. Little wonder that the first naturalists to see the display grounds of the New Guinea bowerbirds were convinced that they must have been fashioned by human hands.

The characteristic feature of the group is the courtship behaviour of the males of certain species which construct arenas in the form of bowers, maypoles, avenues and the like in order to attract the females and encourage them to mate. In the case of the most primitive species the construction is rudimentary; with more highly evolved species it is astonishingly complex and sophisticated.

Not all the bowerbirds are builders. Some males simply court the females from the branches of trees, like other birds. Some, a little more specialised, clear a space on the forest floor, resembling the stage used for a similar purpose by some birds of paradise. A third group decorates such a display ground with miscellaneous objects, often brightly coloured, such as pebbles, bits of glass, snail shells, seeds, fruit and flowers. Finally there is the group which builds elaborate constructions of interwoven grass, in some cases even using leaves or bits of bark as tools for daubing the inner walls with coloured pigments. It must be pointed out that the bowers are unconnected with the nests proper, the latter being rudimentary affairs built in trees by the females.

Sexual dimorphism varies greatly, according to species. Gilliard has made some valuable observations on this point which tend to show that there is an inverse relationship between the colour of the male's plumage and the complexity of his bower. Thus it is the most brilliant and eye-catching male who will build the simplest and least ornamental bower. But among species where there is hardly any outward difference between male and female the bower is likely to be highly elaborate and lavishly adorned with sparkling objects. The inference must be that a drably coloured male compensates for this in the female's eyes by building a spectacular structure which cannot fail to attract her. Brightly coloured birds evidently have no need to make use of external objects to win their mates.

There has been much speculation as to the purpose of these bowers and avenues, and the ornithologist Professor A. J. Marshall has provided an interesting answer. The male apparently begins building his bower as soon as he is mature but only maintains it in a good state of repair and decorates it in the months when he is sexually active. Castration inhibits the building aptitude while hormone injections stimulate it. The female is sexually active for a much shorter period. Only when sexually receptive will she be sufficiently attracted by the bower and its decoration to enter and consent to mate. This does not happen at any fixed time but depends to a large extent on weather conditions being good enough to provide adequate quantities of arthropods – main constituent of the birds' diet. So the male simply has to wait until the female is ready to be courted.

The ground-building bowerbirds may be divided into three principal groups. Those birds that build simple stages chiefly belong to the genera *Archboldia* and *Ailuroedus;* the bowerbirds that construct elaborate bowers and maypoles out of interlaced scraps of vegetation include the genera *Prionodura* and *Amblyornis;* and the birds that lay out avenues and whose bowers are decorated or painted to a lesser or greater degree are representatives of the genera *Ptilonorhynchus, Chlamydera* and *Sericulus*.

The courtship rituals of the first two groups, which live in the rain forests of New Guinea, are little known, with one exception – the Newton's bowerbird (*Prionodura newtoniana*) which, because it is often seen in clearings, has been studied by ornithologists in some considerable detail.

The satin bowerbird builds a display ground which is decorated with matching feathers, shells and other bright objects. He also daubs paint, consisting of pigment mixed with saliva, on the walls of the bower, holding a wad of bark in his bill as a tool.

Regent bowerbird
(*Sericulus chrysocephalus*)

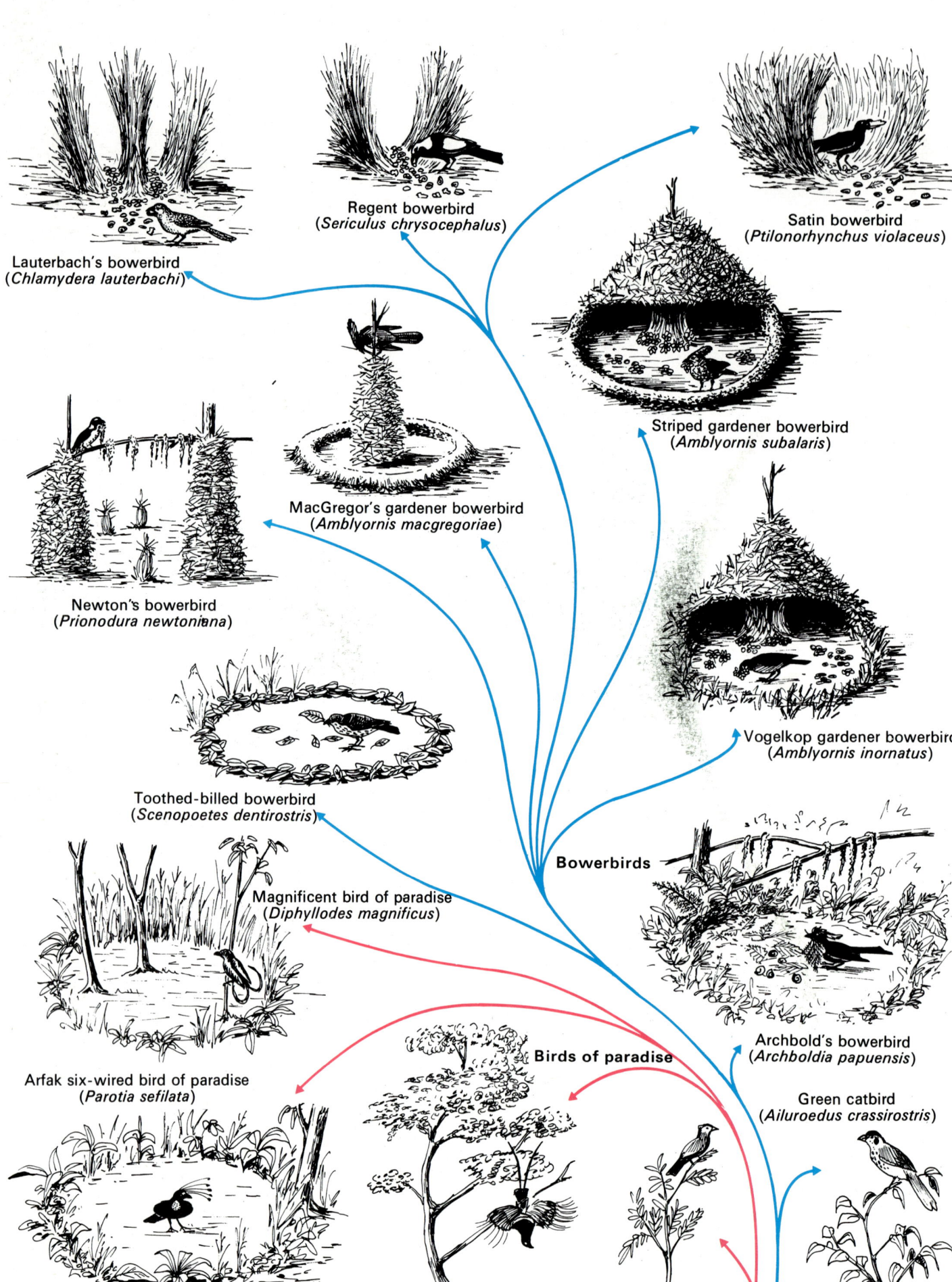

Having completed his structure of stems and twigs, whether it be a dome-shaped bower or an avenue flanked by low hedges and then decorated with miscellaneous objects or touches of 'paint', the male proceeds to clear a space for the courtship display proper. This consists of rhythmic dancing steps, accompanied by trilling calls which sometimes mimic those of other birds, whilst holding a bright object in his bill.

The bowerbirds' breeding behaviour differs markedly from that of other Passeriformes in which both birds normally collaborate in nest-building, preparatory to mating and egg-laying. The very act of building a nest is in these cases an outward expression of mature sexual instincts and the urge to mate. Things proceed otherwise with the bowerbirds. Since the male and female periods and peaks of sexual activity do not coincide, the former occupies the time until his mate shows signs of receptivity by devoting himself assiduously to the preparation and ornamentation of his bower. The female, although putting in an occasional 'goodwill' appearance, will only venture inside the nuptial territory when she is ready to be courted. Until this happens her attentions are chiefly concentrated on building a traditional type of nest some distance away where she will eventually lay her eggs and rear her brood.

There is a good reason to think that as a consequence of the remarkable change of behaviour which urges the males to transform themselves into architects at this crucial time, natural selection, over a period of many thousands of years, has driven these species to undertake ever more complicated feats of construction, eventually producing the highly elaborate display grounds that are nowadays in evidence. Because the females of the various species have clearly responded to the stimuli of

The male bowerbird uses branches, twigs and leaves to construct the mound and avenue where he will later perform his courtship display before the female of his choice. Often he will decorate his display ground with brightly coloured objects (*left*).

*Facing page:* Birds of paradise and bowerbirds are related, being descended from a common ancestor, but have evolved widely differing forms of courtship and mating behaviour. The birds of paradise concentrate their nuptial display on the exhibition of their splendid plumage and unusually shaped crests and tails. Some species perform from a branch, others clear a space in a similar manner to the bowerbirds. The courtship activities of the latter are characterised by the building of grassy and leafy structures, or bowers of sticks, sometimes decorated or painted, the more elaborate bowers being built by the birds with plainer plumage.

bowers and avenues rather than to the personal attractions of the males, the latter have tended to engage in a form of rivalry in their building activities. It is noticeable that the females are most strongly drawn towards the most lavishly adorned bowers. But if beautiful ornamentation is a decisive factor in the eyes of the female it is no less important for the individual male as he stakes out his territory and asserts himself among potential rivals. It is therefore with this dual purpose—to attract a mate of the same species and to ward off other males—that a bowerbird will surround himself with decorative materials that echo the colours of his own plumage, his eyes or the exposed areas of his body. Thus the satin bowerbird (*Ptilonorhynchus violaceus*), whose splendid blue feathers match the colour of his eyes, not only selects blue objects to decorate his display ground but also paints the walls of his bower blue. There is surely no other comparable example in the bird kingdom of the art of 'make-up', if this be defined as the use of matching artificial aids to heighten natural beauty!

Some ornithologists have likened the nuptial dance of the male bowerbird, during which he grips a glittering object in his bill, to a variant of the ritual food offering to the female as exhibited by the males of other passerines. In fact such offerings are common features of the courtship behaviour of most birds. Thus the display of a snail shell or a piece of mother-of-pearl in the mouth of a male bowerbird may well have its origin in a more traditional form of courtship, such as the male offering a twig or a leaf to his mate.

## The ingenious mound-builders

The Megapodiidae or mound-builders are extraordinary fowl-like birds belonging to the order Galliformes, remarkable, above all, for the unusual manner in which they incubate their eggs. It is a small family, widely agreed to comprise seven genera and twelve species which range from eastern Indonesia, through the Polynesian island chain and south to New Guinea and Australia.

There has been some confusion in classification, particularly concerning the scrub-fowl of the genus *Megapodius*. At the beginning of the present century ornithologists identified nine species and twenty-eight subspecies. This has now been amended by Mayr and Amadon who have shown that there are only three distinct species of this genus.

On this basis the Megapodiidae may be quite simply divided, according to habitat, into three groups. In the first are the three species of *Megapodius*, together with Wallace's junglefowl (*Eulipoa wallacei*) and the Celebes maleo (*Megacephalon maleo*); in the second group are the three species of the genus *Talegalla*, the brush turkey (*Alectura lathami*) and the two species of the genus *Aepypodius*. Finally there is the mallee fowl (*Leipoa ocellata*), already described in the chapter on the fauna of the Australian desert. The birds belonging to the first and second groups are typical inhabitants of dense tropical rain forest.

More specifically, the brush turkey's range of distribution is confined to the east coast of Australia while the junglefowl

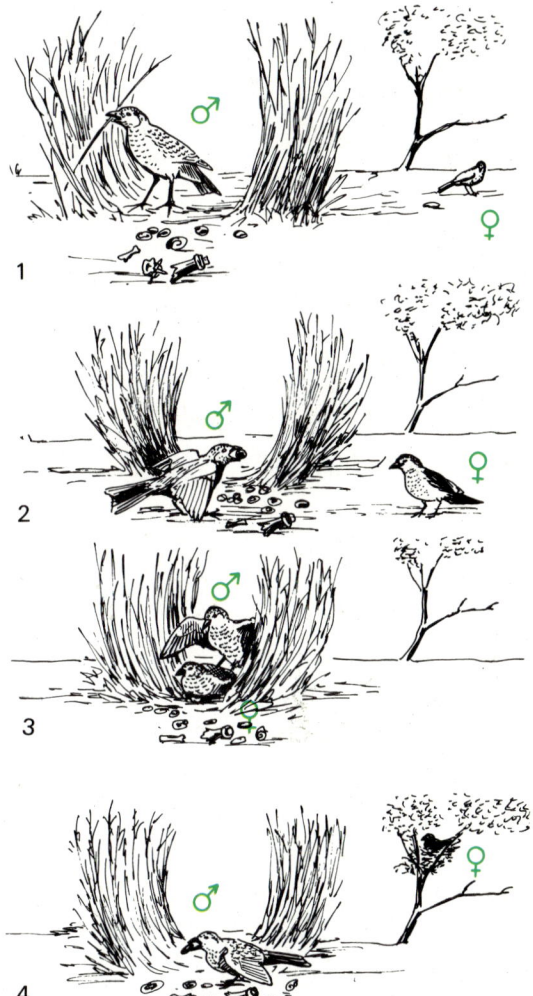

The male bowerbird spends many months building and decorating his bower while the female, not yet sexually receptive, pays little attention to his activities (1). Eventually she shows interest in the nuptial territory and approaches the male who has by now completed his work (2). In the course of his courtship display he picks up various bright objects and exhibits them to the female. Mating then takes place in the bower (3). The female then leaves the display ground and lays her eggs and raises her brood in a nest she has built in a nearby tree. Meanwhile the male continues to renovate and adorn his bower (4).

*Facing page:* The satin bowerbird builds a fairly simple avenue of twigs and grass, with added decorative materials and 'paint' to match the plumage. Although not as spectacular as that of other species the construction is intricate and the entrance just large enough to get through without the bird touching the sides with its wings.

Although all megapodes incubate their eggs artificially the site and structure of the incubation chamber varies according to species. The simplest is that of the Celebes maleo fowl (1) which lays its eggs in a cleft of volcanic rock or in a hole scooped in soil warmed by volcanic gases. The brush turkey (2) builds a mound composed entirely of leaves and other vegetation. The mallee fowl (3) constructs an even more elaborate mound made up both of sand and heaps of rotting vegetation.

*Megapodius freycinet* is essentially a bird of the tropical rain forest.

The megapodes of the first group, especially Freycinet's junglefowl, have a much wider distribution than the others, with a range extending from the Nicobar Islands in the west to Polynesia in the east and the Philippines in the north, and southward to the tropical shores of northern Australia. The species inhabits both rain forests and coral islands where the vegetation is sparser. Wallace's junglefowl and the Celebes maleo fowl spend the greater part of the year in dense forest but during the breeding season they are often seen near the coasts.

As previously mentioned, the mallee fowl is the only representative of the family which lives in desert and semi-desert regions of Australia, and its name is derived, of course, from the various dwarf forms of eucalyptus which grow in this habitat.

All the megapodes are ground birds, constantly on the move. Like most Galliformes they are inexpert fliers and when threatened prefer to run away or to hide in the undergrowth.

Apart from Wallace's junglefowl and the maleo, these birds are not distinguished by vividly coloured or patterned plumage, their feathers being predominantly brownish-yellow with greyish undertones. A few longer feathers on the head form a modest crest. The maleo, however, has black upper parts, with the lower tail coverts and underparts ranging from white to pinkish-brown, and is also distinguished by a curious head casque.

All the forest species of megapodes (*Megapodius*, *Megacephalon* and *Eulipoa*) are approximately the same size, with a total length of about 20 inches. The megapodes belonging to the second group (*Alectura*, *Aepypodius* and *Talegalla*) are somewhat larger, around the 28–30-inch mark. They are all similar in colour, black tending to predominate. The naked parts of the head and neck are covered with coarse down and there is a ruff of yellow or whitish feathers around the neck.

Like the majority of Galliformes, the mound-builders are primarily vegetarians, although their diet is extremely wide-ranging. They consume enormous quantities of seeds, fruit and leaves, supplemented at intervals by insects and small vertebrates. All the representatives of the family have the same type of short, thick bill, the upper mandible being slightly hooked and projecting beyond the lower one. But the most distinctive anatomical feature of these birds is the large size of the toes, disproportionate to the rest of the body. It is from the huge feet, of course, that the family name derives (Gr. *Mega* = large, *Pous* = foot).

As noted in the section devoted to the mallee fowl, the most remarkable fact about the megapodes is the original manner in which they incubate their eggs. The heat necessary for the development of the embryo inside the egg is, among the majority of birds, generated by the body of one or other parent. But in the case of the megapodes such heat is derived by other means – usually either from the rays of the sun or from heaps of decaying vegetable matter, the temperature varying or being regulated according to time of day and season. None of the megapodes incubates eggs in the traditional manner, but the special incubation chambers constructed by the various species differ in interesting ways. The information contained in the following section is largely based on the work of Dr. H. J. Frith, the leading authority in this field.

The brush turkey is one of the largest of Australasian megapodes, an inhabitant of eastern coastal regions of Australia.

## Holes and mounds

Of all the methods of nest building and artificial incubation employed by the Megapodiidae, the simplest are those of Wallace's junglefowl, the maleo and some Freycinet's junglefowl. These species merely deposit their eggs in a shallow depression dug in the earth in the full glare of the sun. The hole is then covered up and once this is done male and female retire, paying no further attention to eggs or brood and making no effort to regulate the temperature inside. But despite the fact that the parents never return to the nesting site, things do not appear to be left to pure chance. Studies on the subject have shown that the birds carry out a careful preliminary examination of the terrain, scooping out their holes in places where the temperature is exactly right. It is interesting to note, for example, that in

Mallee fowl
(*Leipoa ocellata*)

some volcanic areas, as in parts of the Solomon Islands, the holes are deliberately burrowed in places where the hot volcanic gases rising from the ground can be relied upon to provide a stable temperature for incubation.

On the coast of Queensland some megapodes lay their eggs in rock clefts exposed to the sun's rays. The heat absorbed by the stone during the day is sufficient to provide enough warmth by night to keep the incubation temperature fairly constant. In other types of habitat certain Freycinet's junglefowl build enormous mounds that may be 5 feet in height and 40 feet in diameter, simply by piling up earth and dead leaves. As the leaves decompose within the mound sufficient heat is generated to ensure that the eggs will hatch. The eggs are deposited in tunnels leading off in different directions from the centre of the mound, each of them about 3 feet long.

The proportion of soil to vegetation varies according to the place where the nest is built. Some mounds are made up almost entirely of earth whereas others are composed exclusively of gigantic heaps of leaves.

Information so far obtained suggests that among the species concerned several pairs of birds may work together to construct such a nest and that the precise proportion of soil and vegetation is determined by the air temperature and the duration of sunlight hours. Thus in places where there is insufficient sun there will be a larger quantity of accumulated leaves to furnish additional heat.

The brush turkeys of the genus *Alectura,* which are among the largest of megapodes, build mounds composed almost entirely of dead leaves. These measure up to 3 feet high and 10 feet across. In the tropical rain forest the vegetation decays rapidly and the heat thereby released compels the males to keep a constant check on the interior temperature by probing the mound with their bill at regular intervals. In Dr Frith's opinion it is the tongue which functions as a temperature-regulating apparatus. We have seen that the male mallee fowl tests mound temperature in the same way.

When the leaves begin to decompose the temperature is very high and in order to prevent it increasing out of control the male brush turkey tramples the mound and allows air to circulate by scooping out openings here and there and raking the leaves over with his feet. As long as the temperature is not exactly right he prevents the female from laying her eggs inside. During the entire lengthy period of incubation the male continues to test the interior of the mound with the bill; but although the general principle is clear it is not known precisely how he manages to keep the temperature sufficiently constant to insure incubation.

As is the case with the mallee fowl, the female brush turkey lays her eggs at intervals over a period of several months, and since incubation commences immediately each egg is deposited inside the mound the chicks are not born simultaneously but one after another. Unassisted by their parents they are sturdy enough when they hatch to force a path out of the incubation mound and into the fresh air. Once outside they use their legs

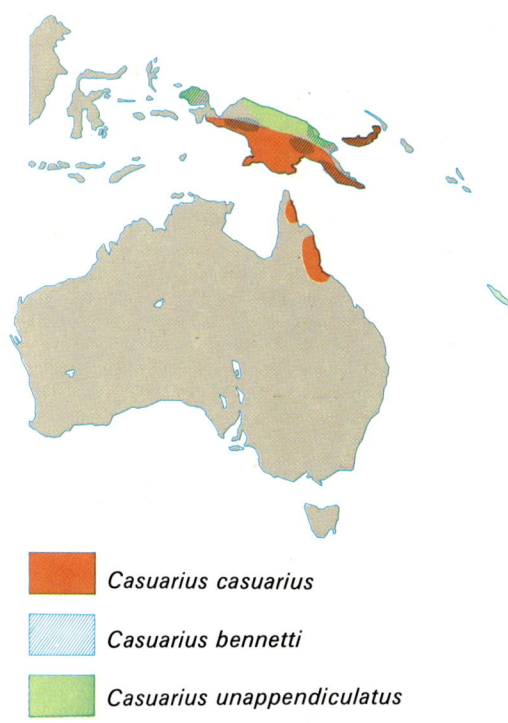

*Casuarius casuarius*
*Casuarius bennetti*
*Casuarius unappendiculatus*

Geographical distribution of cassowaries.

**COMMON CASSOWARY**
(*Casuarius casuarius*)

Class: Aves
Order: Casuariiformes
Family: Casuariidae
Height to shoulder: up to 36 inches (90 cm)
Length of tarsus and metatarsus: up to 12 inches (30 cm)
Diet: basically vegetation but also small animals
Number of eggs: 3–6
Incubation: 49–56 days

Black, drooping plumage. Primary feathers vaneless, like wiry quills; wings rudimentary, unsuited for flying. Large casque on forehead; upper part of neck naked, dark blue; two large red wattles hanging from front of throat. No external difference between male and female, but latter slightly larger. Other species are Bennett's cassowary (*Casuarius bennetti*), smaller and without wattles, and the single-wattled cassowary (*Casuarius unappendiculatus*), larger, with one wattle in centre of throat.

*Facing page:* The blue-crowned pigeon is one of three species making up a family of Columbiformes found only in New Guinea. It is a large ground bird, widely hunted.

to scamper off into the undergrowth. Within twenty-four hours they are capable of flying. All Galliformes are notable for the rapid development of their young, but in no other family are the babies so completely independent and self-reliant as these precocious chicks. At no stage in their life do they even come into contact with their parents so that in the most literal sense preparation for their solitary existence begins on the very day they are born.

## The forest-dwelling cassowary

The cassowaries, together with the emus, are large ratites belonging to the order Casuariiformes. Thus they are only distantly related to the ostriches (Struthioniformes) and the rheas (Rheiformes) despite some obvious resemblances. Cassowaries are further distinguished from emus by certain external characteristics as well as by habitat. Consequently they have been placed in separate families. The emu, of which there is today only one species, belongs to the Dromaiidae. Other species were once found in Tasmania and small neighbouring islands off the south coast of Australia but these are now extinct. Larger than the cassowary, the emu has feathers on the head and neck, apart from two naked patches on either side of the head, but lacks the casque, vaneless primary feathers and spiky inner toe of the cassowary. It lives on the arid and semi-arid plains of Australia. The three species of cassowary (Casuariidae) have a horny casque on the forehead, distinctive naked coloured patches on face and neck, and sometimes drooping wattles. They are inhabitants of the rain forests of northern Australia, New Guinea and neighbouring islands. Neither family is to be found outside the Australasian region.

The Australian ratites may easily be distinguished from related species in other parts of the world by the fact that they have no tail feather or uropygial (preen) gland (found at the root of the tail in most birds and secreting an oily substance used in preening). Furthermore they have only six or seven primary feathers, the wings being rudimentary—so undeveloped in fact that the combined length of the ulna, radius, carpus, metacarpus and phalanges is equal to that of the humerus. The green eggs have pitted, very rough shells.

Apart from these detailed variations the Australian flightless birds are immediately recognisable by reason of a longer body, shorter legs and a less developed neck which for half its length is covered with feathers.

There are today three species of cassowary. The best known of these is the common cassowary (*Casuarius casuarius*), measuring up to 3 feet at the shoulder. Its convex head casque is larger than that of related species and a pair of red wattles hang from the throat. The upper part of the neck is dark blue, the skin being roughened by deep wrinkles. The common cassowary has a somewhat wider range than the others—the southern half of New Guinea and adjacent islands, as well as the Cape York peninsula of northern Queensland. Ornithologists have identified eight subspecies of common cassowary.

The cassowary is a strongly built ratite with powerful, stumpy legs. The common cassowary, shown here, has a large, rounded head casque and a pair of wattles hanging from the throat.

The single-wattled cassowary (*Casuarius unappendiculatus*) is the largest representative of the family, standing well over 3 feet at the shoulder. The casque is flattened at the rear, and the neck, which is blue, is encircled half-way down by a red ring of feathers covering the sides of the lower, naked neck areas. As its name suggests, this species possesses only a single wattle which hangs from the centre of the lower throat. In addition there are two fleshy growths at the base of the bill. This bird is an inhabitant of northern New Guinea and a couple of nearby islands. Four separate geographical subspecies have been identified.

Smallest member of the genus is Bennett's cassowary (*Casuarius bennetti*), which stands approximately $2\frac{1}{2}$ feet at the shoulder. The rounded casque on its head is not greatly developed and it lacks wattles. There are large blue and orange naked patches on the neck. This bird is found in the western and eastern regions of New Guinea and in New Britain. Seven geographical races have been identified, one of which, the Papuan cassowary, is sharply differentiated from the others in several respects.

## The kick that can kill

All the cassowaries are powerfully built birds which may weigh up to about 150 lb. In contrast to other ratites the females are normally a little larger than the males. Although smaller than ostriches, they are a good deal stronger and a well-aimed blow from one of their feet (a jumping action reinforced by jabs with

The cassowary, unlike the emu, is a bird of the forest. It is a rapid runner and an expert swimmer. This is the single-wattled cassowary.

Common cassowary
(*Casuarius casuarius*)

The cassowary is a dangerous adversary when cornered, jumping at an enemy with flailing feet. The sharp claws, particularly the stiletto-like claw of the innermost toe, are capable of inflicting serious, often fatal, injuries.

*Facing page:* The hard, spiny feathers of the cassowary are said, on little evidence, to protect the body as the bird brushes against thorns and branches in its forest environment. The casque too is used for pushing aside natural obstacles.

the sharp claws) is capable of killing a man.

Apart from the kiwis, cassowaries are the only flightless birds living in dense forest. They are unable to survive outside this habitat and are usually to be found close to watercourses, as on the banks of lakes and rivers. The strange casque or helmet which is such a distinctive feature of the head has a definite purpose in the forest, for the bird will use it as a kind of cutting tool to hack out a path through dense undergrowth.

There are several interesting features of the coarse, drooping plumage. Each feather is double, in the sense that a secondary feather, almost as long as the main one, sprouts from the base of each shaft. The short, rudimentary wing feathers lack barbs proper but bear four or five rounded, vaneless quills, rather like horny spines. These long, curving quills are often said to have a function too, protecting the body against thorns and other obstacles as the bird bounds through the forest. But some scientists consider this unlikely.

Cassowaries feed mainly on vegetation, but like the ratites of the open plains which frequently supplement their diet with arthropods and small vertebrates. Despite their size they emulate other animals of the forest in being wary and silent by nature, so much so that it is almost impossible to locate them and hence study them in their wild surroundings. If surprised and alarmed they will make off at top speed. Although there may be all manner of obstacles in their path they will bound along at up to 30 miles per hour, head downward so that the body is in a straight line (unlike ostriches and rheas which run with head held upright), pushing aside everything in their way. It is amazing to see how effortlessly such large birds can stride through the thickets, leaping clear over bushes and hillocks when the need arises. Even a body of water will not necessarily halt their headlong flight for they are good swimmers and experience little difficulty in fording a river or lake. This may seem remarkable in such a long-legged bird except that others, for example individual herons, will sometimes habitually swim.

If attacked and at bay an adult cassowary is a formidable adversary indeed, beating its wings and kicking out at the enemy with its powerful feet. The latter are particularly dangerous, even deadly, weapons. Two of the three toes bear long, stout claws but the innermost toe is equipped with an even longer claw which is pointed like a spike and capable of inflicting a deep, perhaps fatal, wound. Even if the claws fail to find their mark the impact of the kick itself may be enough to break a bone.

## Breeding problems

In certain respects cassowaries do resemble other ratites. The male of the various species, for example, is responsible for a range of domestic duties, including incubation. The 3–6 enormous, rough-grained, dark green eggs (each of which weighs almost $1\frac{1}{2}$ lb) are deposited by the female in a hastily prepared nest of leaves on the ground and then it is the male who takes over. Very often he will also play a leading role in protecting and rearing the chicks, perhaps spending 7–8 weeks in looking after his

Ratites of the open plains, such as the ostrich and the rhea, run with the head held rigidly upright so as to obtain a clear view in all directions. The cassowary, however, like other forest animals, runs equally fast but with neck and head held horizontally to guard against low-lying obstructions.

Ostrich · Cassowary

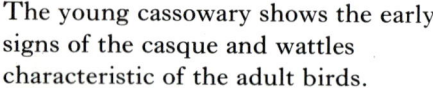
The young cassowary shows the early signs of the casque and wattles characteristic of the adult birds.

progeny. As with other ratites the young are nidifugous, with nondescript plumage that matches the surroundings to perfection and allows them to skip unseen into the bushes, hopefully escaping the teeth and claws of predatory mammals, birds and reptiles. The adult plumage too blends with the forest environment and the first instinctive action when danger threatens the nest is for the bird to flatten itself against the ground in the hope of remaining undetected. If this ruse fails it will jump up and run, perhaps carrying out a series of diversionary manoeuvres to lead the enemy away from the nest. These manoeuvres, known as distraction displays are an especial feature of all ground-nesting birds. They are particularly elaborate in the ratites.

Once adult, and apart from the times when they have a family to raise, cassowaries have no natural enemies, except for man. They have long been hunted by the Aborigines for their tasty flesh and in particular their large liver. Because they are the only animals, together with wild boars, that are regularly hunted because of their food value, the Aborigines of New Guinea and northern Australia treat them with familiarity and in many villages it is common to see young birds wandering freely among the huts. But when they grow to adulthood they have to be treated with much more care and their owners place them in enclosures from which they are released only to be eaten or sold. Apart from their flesh, cassowaries are much prized for their feathers and there is a brisk local trade in the birds for both these reasons. Sometimes their owners will take them, either on foot or in boats, to be sold in distant villages—a confusing situation for the ornithologist trying to trace the precise geographical origin of the various known subspecies. In Papua, according to Gilliard, a cassowary is highly valued at the equivalent of either eight pigs or one woman!

Although ostriches, rheas and emus reproduce without much difficulty in captivity there are problems involved in getting cassowaries to breed in zoos and there are very few places where such experiments have succeeded. The first attempt was at London Zoo in the mid-19th century, but although a chick was hatched it died shortly afterwards. Since then only San Diego, Frankfurt and Dresden Zoos have managed to breed cassowary chicks. Apart from other difficulties related to feeding and rearing, one basic problem, strange as it may seem, is to tell the sexes apart. There is little external distinction between male and female and even when partnered the difficulties are not over, for these solitary birds are by temperament so suspicious and

aggressive that instead of mating they may attack and even kill each other. It is hardly surprising, therefore, that a zoo director may think twice before conducting a breeding experiment in which he risks losing one of his rare birds.

Dr Sanf, who has made a detailed study of the biology and behaviour of cassowaries, states that the first such bird to be seen in Europe arrived in Amsterdam in 1597 and that it was eventually offered as a gift to the Holy Roman Emperor Rudolf II. After that, and especially during the early years of the present century, the birds were introduced to many of the world's zoological gardens. But although they have been studied in captivity it has proved difficult, for reasons already explained, to discover much about their habits in the wild. Clearly they live a solitary existence, only congregating in small mixed groups during the breeding season. Research on captive cassowaries suggests that incubation lasts 49–56 days.

The common cassowary and the single-wattled cassowary are inhabitants of the flatland forests but Bennett's cassowary is also sometimes found up to a height of 10,000 feet.

Although there is little outward difference between the adult male and female cassowary, the young bird, as can be seen in this picture, lacks the thick black plumage of the adult as well as the vivid neck and head coloration.

Tawny frogmouth
(*Podargus strigoides*)

Although trading in animals is not normally to be condoned, it is evidently as a result of such transactions at local level that cassowaries have spread well beyond their original habitats, so that in a sense the birds may be regarded, in certain areas, as introduced species which have reverted to the wild. Such individuals, incidentally, are notable for their size.

Taxonomic study of the Casuariidae is not a simple matter for there is much variation among birds of the same species, let alone the geographical problems already mentioned. Some authors distinguish species which in the opinion of other experts are mere races. But it is widely accepted that there are in fact no more than three distinct species, as described in the preceding pages.

## The tawny frogmouth

The frogmouths of the Australasian region belong to the order Caprimulgiformes and are thus related to the widely distributed nightjars of the Old and New Worlds. The family Podargidae, in which they are classified, is made up of two genera and twelve species, ranging from the eastern Himalayas to Papua, Australia and Tasmania.

So named because of their enormous mouth, they are distinguished from European counterparts by an extraordinarily strong, thick bill. The typical Australian species, the tawny frogmouth (*Podargus strigoides*), is a tree-dwelling bird, constructing a large nest in the fork of two branches, in which the female lays two or three white eggs. Nocturnal by habit, it sleeps by day perched lengthwise along a branch, head only slightly raised. This characteristic posture, together with the nondescript matching plumage, disguise the bird most effectively, so much so that it looks like a piece of dry branch.

The tawny frogmouth, like its relatives, is a clumsy flier and instead of catching insects on the wing it preys on crawling insects as well as on fledglings and small mammals on the ground. But because the species is only active by night relatively little is known of its habits.

Among the anatomical features distinguishing frogmouths from nightjars of the family Caprimulgidae are the powdery down on the back and the presence of thirteen instead of fourteen cervical vertebrae.

## The marvellous lyrebirds

Two-thirds of the world's birds (more than 5,000 species) belong to the order Passeriformes, commonly known as perching birds. The order is subdivided into four suborders, each made up of families that show much diversity of form. The Eurylaimi comprise only one Old World family, the broadbills. The Tyranni of the New World contain thirteen families, including ovenbirds and tyrant flycatchers. Four-fifths of the world's Passeriformes belong to the suborder Passeres, or true singing birds; and the Menurae, which are peculiar to Australia, consist of two families, the lyrebirds (Menuridae) and scrub-birds (Atrichornithidae),

*Facing page:* The tawny frogmouth is a nocturnal forest bird whose plumage blends well with the branch on which it habitually perches. Unlike related species it takes most of its prey on the ground instead of catching it on the wing.

The male superb lyrebird raises its tail to a vertical position during the courtship display but only for a few moments, the more typical posture being horizontal, over the back.

each with a single genus and two species. The former is the better known family and it is the lyrebird which appears on Australian postage stamps and seals.

Australia is a continent which specialises in fascinating animals, curious both in appearance and behaviour. The kangaroo, the duckbill or platypus and the echidna, the emu and the cassowary are all distinguished in their own ways, either for their style of locomotion or their unusual breeding habits. But for sheer beauty and originality of form the leading contender is surely that marvel of nature, the lyrebird.

According to Thomas Gilliard, the great authority on Australian birdlife, it was in February 1798 that a large bird, assumed to be a form of pheasant, was caught in a mountain forest of New South Wales. It was immediately evident that this was a scientific discovery of considerable importance. European ornithologists who now had the opportunity of studying it for the first time eventually named it the lyrebird.

Although it was at first taken to be an ornamental species of pheasant this large bird was duly classified, correctly, as one of the Passeriformes and thus a distant relative, improbable though it may seem, of the common sparrow. The two recognised species are the superb lyrebird (*Menura novaehollandiae*) and Prince Albert's lyrebird (*Menura alberti*), named after Queen Victoria's Prince Consort.

The range of the superb lyrebird extends down a narrow belt of the eastern Australian coastline. Albert's lyrebird has a much smaller, more northerly distribution.

In external appearance the Menuridae somewhat resemble the megapodes but they are distinguished from the latter by their primitive oscine type of syrinx (vocal organ) whose structure is in certain respects similar to that of songbirds. Moreover, the breastbone of the lyrebird is long and narrow, not squarish like that of other passerines and closer, in fact, to that of most water birds.

The head is small in relation to the rest of the body but the bill is comparatively large. The legs are long and sturdy, the feet being furnished with long toes and powerful, curved claws. This last feature, together with the brownish colour of the plumage and the length of the tail, initially led scientists to suppose that it belonged to the Galliformes.

The lyrebird's tail is unique, consisting of sixteen feathers instead of the customary twelve. In the male superb lyrebird the two thick outer feathers, each close to 2 feet in length, form the framework of what has been likened to an ancient lyre. These feathers are whitish with prominent brown V-shaped patterns, curving outwards in the graceful shape of a crescent. The next six pairs of feathers are long and filmy with separate vanes; and the central pair are unadorned, wire-like plumes sweeping beyond the tips of the lyre.

Of the two species the superb lyrebird is unquestionably the more spectacular and it is the male, his tail outspread for the courtship display, which features on stamps and seals. His total length is slightly over 3 feet whereas Albert's lyrebird is about 6 inches shorter. The tail of the latter is less showy.

Although roughly shaped like a lyre, the outer feathers are not so boldly patterned and do not form the frame. Furthermore, the central rectrices have a normal type of web. The colours too are generally more rufous and less striking.

The superb lyrebird is well accustomed to human presence and is frequently seen in the bushland near the larger Australian cities. Albert's lyrebird, however, is more rarely encountered for it is an inhabitant of dense forest, well off the beaten track, and little is known of its habits.

In both species there is marked sexual dimorphism. The females are much smaller, with a similar plumage, but their tails are of ordinary shape.

The lyrebird is renowned for its astonishing repertory of

The male lyrebird commences his courtship performance on a low branch and then continues it on the ground, often standing in a hollow or on a mound. The tail is spread to reveal its lyre-like shape, held vertically and then brought forward to cover the back and the lowered head. The display also includes hopping movements and a varied exhibition of the bird's vocal powers.

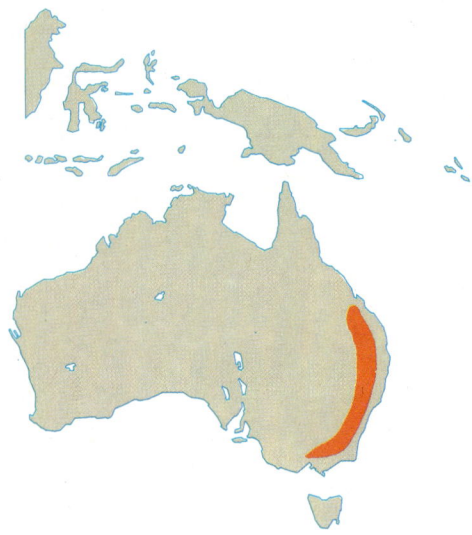

Geographical distribution of the superb lyrebird.

---

**SUPERB LYREBIRD**
(*Menura novaehollandiae*)

Class: Aves
Order: Passeriformes
Family: Menuridae
Total length: up to about 40 inches (100 cm)
Diet: mainly vegetation
Number of eggs: one
Incubation: about six or seven weeks

Colour greyish-brown. Very long tail, in shape of lyre, with large outer feathers forming frame and inner feathers filmy, resembling strings of the instrument. Female has shorter tail. Nidiculous chick born almost naked and helpless.

---

*Facing page:* Sexual dimorphism is very noticeable in the superb lyrebird. The female (*above*) is smaller and the tail, although long, lacks ornamental features. The male (*below*) is distinguished by its lyre-shaped tail, comprised of sixteen feathers, the two outer ones being boldly patterned and providing a frame for the others.

vocal sounds. There is hardly any other bird species to rival it in this respect. In addition to an extensive range of notes in varying registers, the bird mimics other sounds, including birdsong, with astounding accuracy. Mingled with the raucous laughter of the kookaburra and the screech of the cockatoo will be the mewing of a cat, the barking of a dog, the bleat of a sheep, the buzz of machinery, the honking of a car horn and a miscellany of other noises, both natural and artificial, according to the surroundings. Apart from its astonishing flexibility the lyrebird's voice is remarkably penetrating, audible in sparse woodland up to a mile away.

The courtship display of the male superb lyrebird is a truly unforgettable sight and certainly ranks as one of the most astonishing performances to be witnessed in the entire bird kingdom. Having selected his piece of display ground the male piles up a small mound out of branches, twigs and dead leaves or alternatively scoops out a hollow, either in a clearing or in the grass. Whatever guise it takes, this stage is used for the elaborate song and dance ritual which precedes mating.

Unlike other species whose sexual encounters take place on clearly defined nuptial territory, the cock lyrebird is monogamous. When a female lyrebird tentatively approaches the display ground the male, having warbled a greeting from a tree stump or a low branch, flutters down and initiates the courtship performance proper. While he sings he slowly begins to extend his magnificent tail, in the same manner as a peacock or a megapode in similar circumstances. For a few fleeting seconds only he raises the tail vertically to reveal the lyre shape in all its glory.

Ornithologists studying this ritual were for a long time in two minds as to whether the male lyrebird did in fact raise his tail in this manner; but recent observations have shown conclusively that it is indeed maintained in this upright position, even if only for a very brief period. After that it is extended in a horizontal direction at an angle of 180° or more, so that the quivering feathers envelop the back and head like an enormous silvery fan, the tips touching the ground in front. Now, with his head almost invisible beneath the flowing tail, the bird performs a little hopping dance, revolving in tiny circles, trilling and warbling all the while. Suddenly he lets out a series of high, piercing notes which signal the conclusion of the display. The tail is slowly folded back into its customary position and he stalks haughtily away, the female in his wake.

The reason for this detailed discussion of how the male lyrebird holds its tail in display is that it is always pictured, for example on postage stamps, with the tail held vertically in what is now known to be an atypical posture.

It is the female lyrebirds who build the domed nests, on the lower branches of trees, on rock ledges or occasionally high up in a tree. About a week later each bird lays a single egg in the nest. It is approximately the size of that of a domestic hen and is in fact the largest laid by any passerine. It has a thick shell and is greyish with elaborate violet markings. Incubation lasts six or seven weeks and this too is a record for Passeriformes.

The chick is nidiculous, being reared in the nest, and grows

extremely slowly. Always hungry, it will often pierce the domed roof of the nest, thrusting its head outside in its eagerness to be fed. Not until the youngster is two months old will it venture out with hesitant steps in its mother's tracks and begin exploring the undergrowth. Initially both sexes look like an adult female. The male takes on his splendid adult plumage at the age of two to three years.

Mid-autumn is the usual mating season and the females begin their nest-building activities in May or June. This means that the young are born in the winter, a season when they can be fed plentifully on crustaceans, worms and other small animals. A certain quantity of animal protein is clearly necessary for normal development but both species are believed to feed principally on vegetation of various kinds. The exact constituents of their diet are not, however, known.

Lyrebirds, being rapid runners, spend most of their time on the ground although, as has been noted, they occasionally perch in trees. They are not strong fliers, confining their aerial manoeuvres to short glides at low altitude.

The birds are not seen to best advantage in captivity. They sing very little, seldom perform their spectacular courtship dances and in fact rarely breed in zoos.

## The rare scrub-birds

In 1842 John Gilbert discovered a curious bird in Western Australia. The body was dark chestnut, the head and breast flecked with white. Its feet were large and strong and it measured about 8 inches long. Two years later it was named by Gould as the noisy scrub-bird (*Atrichornis clamosus*). Later, in 1865, a similar bird was found in New South Wales, an inch or so smaller than the other, mainly dark brown with reddish-orange throat and underparts. This related species was given the name of rufous scrub-bird (*Atrichornis rufescens*).

Despite their modest dimensions and rather undistinguished looks these birds are considered by some experts to be relatives of the splendid lyrebirds, making up the second family of the passerine suborder Menurae.

Following Gilbert's discovery hardly any further reports were received of the western species and over the years not more than twenty specimens were despatched to various museums in Australia, London and North America. Until 1940 all these museum specimens were believed to be males but then a female was discovered in the Gould Collection in Philadelphia. Since the last bird, however, dated from 1889 it was generally assumed that the noisy scrub-bird was extinct, but in 1961 an isolated male of the species was caught only about twenty miles from the spot where Gilbert had made his original discovery.

During the years that followed a number of other noisy scrub-birds were sighted and the Australian government decided to take urgent steps for the protection of the species. Naturalists have been able to obtain photographs of the birds and their song has been successfully recorded on tape. Recently an empty nest was found, furnishing a little more much-needed information.

This nest was concealed in tall grass a few feet above ground level. It was made up of loosely interlaced dry leaves and the interior was apparently lined with a pulpy substance identified as predigested wood.

Although ornithologists have been unable to record other detailed observations of this rare species, the behaviour of the rufous scrub-bird is far better known for this bird frequents more densely populated areas and is relatively abundant. Nowadays there are two principal centres of distribution, one in the coastal forests of New South Wales, the other in south-western Queensland, these communities being quite separate and over 300 miles distant from each other. Although numerous in comparison with the noisy scrub-bird farther west this species at one time obviously enjoyed a much wider range.

The nest of the rufous scrub-bird is domed—a miniature version of that of the lyrebird—and the interior is plastered with soft wood pulp. The completed nest involves about a month of laborious work on the part of both birds. The female eventually lays two pale pink, brown-flecked eggs.

The Australian black-shouldered kite (*right*) and the letter-winged kite (*left*) are typical raptors of open woodland, preying on large insects and small vertebrates.

The birds feed on insects, worms and snails. They are speedy on the ground, darting silently through the undergrowth, but are poor fliers for the clavicles are vestigial and this seems to impede effective wing movement. Despite their modest size both species of scrub-bird possess astonishingly loud, penetrating voices and are reputedly clever mimics.

Although the shy and secretive scrub-birds appear to be related to the lyrebirds, such links being based on certain apparent similarities of anatomy and reproductive behaviour, not all ornithologists accept these findings.

## Australasia's birds of prey

Although the Australasian region cannot boast anything like the numbers and different species of raptors that are to be found in other parts of the world, those that are resident exhibit a complex range of hunting techniques. Some groups are highly eclectic in their feeding habits, others are more narrowly specialised. Furthermore, these winged predators patrol the skies above every type of terrain, just as they do in other zoogeographical regions. Consequently there is no Australasian animal, be it mammal, bird, reptile, fish, amphibian or invertebrate, which is not in some manner or at some time threatened by one or more of these raptors.

The commonest bird of prey in Australia, with a wide food range, is the kestrel (*Falco cenchroides*), which feeds on grasshoppers and other insects in addition to snails and small vertebrates. It is also often seen in cities, where it nests on tall buildings.

Among other unspecialised hunters brief mention must be made of the fork-tailed kite (*Milvus migrans*), extending its range south-eastwards from Eurasia where it is known as the black kite, the square-tailed kite (*Lophoictinia isura*), the black-breasted buzzard kite (*Hamirostra melanosternon*), the swamp harrier or marsh hawk (*Circus approximans*), the spotted harrier (*Circus assimilis*), the Australian black-shouldered kite (*Elanus notatus*) and the letter-winged kite (*Elanus scriptus*). None of these raptors has a precisely defined habitat and unlike related species in the Old World they tend to prey indiscriminately on any animals they can take unawares, including large insects, small mammals and reptiles, baby birds (especially those of ground-nesting species), as well as feeding on carrion. Obviously such an eclectic diet lacks uniformity and will vary in accordance with seasonal availability, opportunity, size of population and individual appetite. Consequently birds which for a time appear to concentrate almost exclusively on insects, for example, are quite likely to switch suddenly to a diet of snakes or nestlings, depending on what type of prey can most easily be found.

Among the more specialised raptors of the Australian eucalyptus and tropical rain forests are a number of goshawks and sparrowhawks, highly adept at flushing mammals and birds from trees and undergrowth. The most striking of these—and not unlike related European species—are the Australian brown goshawk (*Accipiter fasciatus*) and the white, grey or vinous-chested goshawk (*Accipiter novaehollandiae*), the pale plumage of the

Spot-tailed sparrowhawk (*Accipiter trinotatus*)

Black-mantled sparrowhawk (*Accipiter melanochlamys*)

*Facing page:* Although in comparison with other parts of the world the Australasian region contains a relatively small number of different species, the various diurnal and nocturnal raptors exhibit a wide range of hunting techniques. Many of them, as this diagram shows, have adapted to more than one habitat. Among them are
1. Peregrine falcon (*Falco peregrinus*).
2. Brown hawk (*Falco berigora*). 3. New Guinea harpy (*Harpyopsis novaeguineae*).
4. Osprey (*Pandion haliaetus*). 5. Australian black-shouldered kite (*Elanus notatus*).
6. Fork-tailed (black) kite (*Milvus migrans*). 7. White-breasted sea eagle (*Haliaeetus leucogaster*). 8. Wedge-tailed eagle (*Aquila audax*). 9. Australian kestrel (*Falco cenchroides*). 10. Square-tailed kite (*Lophoictinia isura*). 11. Little eagle (*Hieraaetus morphnoides*). 12. Barn owl (*Tyto alba*). 13. Spotted boobook (*Ninox novaeseelandiae*). 14. Grey goshawk (*Accipiter novaehollandiae*). 15. Swamp harrier (*Circus approximans*). 16. Sooty owl (*Tyto tenebricosa*). 17. Powerful owl (*Ninox strenua*). 18. Black-mantled sparrowhawk (*Accipiter melanochlamys*).

Food preferences of the wedge-tailed eagle.

Oriental hawk owl of South-east Asia (*Ninox scutulata*)

latter being incomparably handsome and occurring in two colour phases, white and pearl grey. Another particularly rare and secretive woodland raptor is the magnificent red goshawk (*Erythrotriorchis radiatus*); and a much smaller species, but one of the most agile of all these birds of prey, is the collared sparrowhawk (*Accipiter cirrhocephalus*).

Specialised hunters of fishes in coastal waters, lakes and rivers include the globally distributed osprey (*Pandion haliaetus*) and the white-breasted sea eagle (*Haliaeetus leucogaster*).

Raptors of more open woodland and plains with scattered tree cover include the crested hawk (*Aviceda subcristata*) and the little eagle (*Hieraaetus morphnoides*), the latter a far more accomplished hunter than its Palearctic relative, the booted eagle (*Hieraaetus pennatus*). In fact in its behaviour and hunting technique it reminds one of a buzzard.

It is worth noting that there are no vultures in the Australasian region and that their ecological niche is here occupied by the wedge-tailed eagle (*Aquila audax*), a bird which feeds principally on carrion but which also hunts live prey in the form of large birds and medium-sized mammals. So in a sense this impressive raptor plays a double role, both as a scavenger and as a super-predator, comparable to the mighty golden eagle of the Holarctic region.

Those characteristic raptors of open plains, the falcons, are

Wedge-tailed eagle
(*Aquila audax*)

When excited or alarmed the wedge-tailed eagle erects the long feathers of head and neck so that its face is framed by a small crest.

also represented in Australian skies, although there are only five species. The impressive black falcon (*Falco subniger*), with its huge wings, is found in the interior of the continent. Another rare bird is the very beautiful grey falcon (*Falco hypoleucos*); and the cosmopolitan peregrine falcon is here represented by the subspecies *Falco peregrinus macropus*. The European lanner falcon also has its counterpart in the shape of the Australian hobby or little falcon (*Falco longipennis*).

This brief and admittedly incomplete catalogue of Australasian birds of prey must conclude with a mention of the very interesting brown hawk (*Falco berigora*), a solitary species with extremely varied feeding habits which is to be found both in woods and over open plains. A peaceful and somewhat indolent bird, it spends the greater part of its time perched on a branch or rock, ready to pounce on an insect, reptile or small bird.

## The nocturnal forest hunters

The haunts best suited to the specialised needs of the nocturnal raptors of the Australasian region are naturally woods and forests, especially the tropical forests of New Guinea and the neighbouring islands, areas which are perpetually plunged in deep shadow. Because there is little appreciable difference between day and night in these dark domains birds of prey that are normally nocturnal can venture out to hunt whenever conditions happen to suit them and can retire to sleep at more or less any time, once their appetites are satisfied.

Among such birds of prey which may be active all round the clock are the so-called hawk owls, consisting of two genera. All but one belong to the genus *Ninox*, among them the rufous owl (*Ninox rufa*), the powerful owl (*Ninox strenua*), the spotted boobook (*Ninox novaeseelandiae*) and the sooty-backed hawk owl (*Ninox theomacha*). All possess rudimentary facial discs, tending to prove that they do not rely exclusively on keen vision for hunting and that hearing plays an equally important role. Some of these species are as large as eagle owls, others as diminutive as little owls. All enjoy a wide food range, their victims including rabbits as well as phalangers and other small marsupials, together with woodland birds.

Similar in all essentials to the adaptable hawk owls of the genus *Ninox* is the sole representative of the genus *Uroglaux*, the New Guinea hawk owl (*Uroglaux dimorpha*). This bird of prey is an inhabitant of the New Guinea tropical forest and feeds on insects, small mammals and birds.

The barn owls, more widely distributed around the world than any other specialised group of birds with the exception of peregrine falcons, are represented in the Australasian region by the common barn owl (*Tyto alba*), the sooty owl (*Tyto tenebricosa*), the grass owl (*Tyto capensis*) and the masked owl (*Tyto novaehollandiae*), the last a large species measuring up to 19 inches in length. All of these nocturnal birds of prey are most adaptable, active both in shadowy forests and above open plains, their habits thus being similar to those of their European counterparts.

Geographical distribution of the grey falcon.

**GREY FALCON**
(*Falco hypoleucos*)

Class: Aves
Order: Falconiformes
Family: Falconidae
Wing-length: male $10\frac{1}{2}$-12 inches
(26.8-30.2 cm)
female $12\frac{1}{2}$-$13\frac{1}{2}$ inches
(31.5-33.8 cm)
Length of tail: male $5\frac{1}{2}$-$7\frac{1}{4}$ inches (13.8-17 cm)
female $6\frac{1}{4}$-$7\frac{1}{2}$ inches
(15.6-18.5 cm)
Diet: small birds and mammals, lizards, insects
Number of eggs: 2-4

Head and back dark grey, tail grey, the tip white with twelve dark bands. Underparts pale grey or whitish with darker markings. Eye chestnut.

Grey falcon (*Falco hypoleucos*)

*Facing page:* The little eagle (*above*) bears a close resemblance to the European eagle. The grey goshawk (*below*) is an inhabitant of eucalyptus woods, clearly related to the European goshawk.

# CHAPTER 100

# The raucous, radiant world of parrots

Few birds have so captured man's fancy and imagination as the parrots and their numerous and diversified relatives. More than 300 species of the order Psittaciformes are distributed throughout the world's tropical regions—an indescribably colourful population with a range that embraces the uncharted recesses of the Amazon jungle, the scattered island chains of the Antilles, the savannah and equatorial forests of Africa and Madagascar, and the woods and forests of both South-east Asia and Australasia.

The two principal centres of distribution are the Amazon basin and the Australasian region, and the claim of the latter to be unrivalled for sheer exuberance of form and colour is supported by the fact that of the six recognised subfamilies, three—the Lorünae, Kakatoinae and Micropsittinae—are exclusive to this zoogeographical region and that two others—the Nestorinae and Strigopinae—are confined to New Zealand. The only subfamily which can really be said to be distributed all over the tropics is the Psittacinae.

Apart from their more obvious attractions (a truly astonishing range of colours, shapes and sizes, coupled with the ability of certain species to mimic the human voice and to memorise and reproduce words or phrases) the parrots and their allies are enormously diversified. Some have specialised food habits, taking seeds, pollen or fruit, others are omnivores and a few of them are carnivores. Certain species are rapid fliers, some are accomplished climbers, others are incapable of leaving the ground at all. In New Zealand there is even a nocturnal species, the kakapo or owl parrot, nowadays so rare as to be in danger of extinction. Yet despite their varied habits and haunts all have certain anatomical features in common, making identification

*Facing page:* The scaly-breasted lorikeet, like the majority of parrots, nests in tree hollows, using its hooked bill to scoop a hole in dead wood.

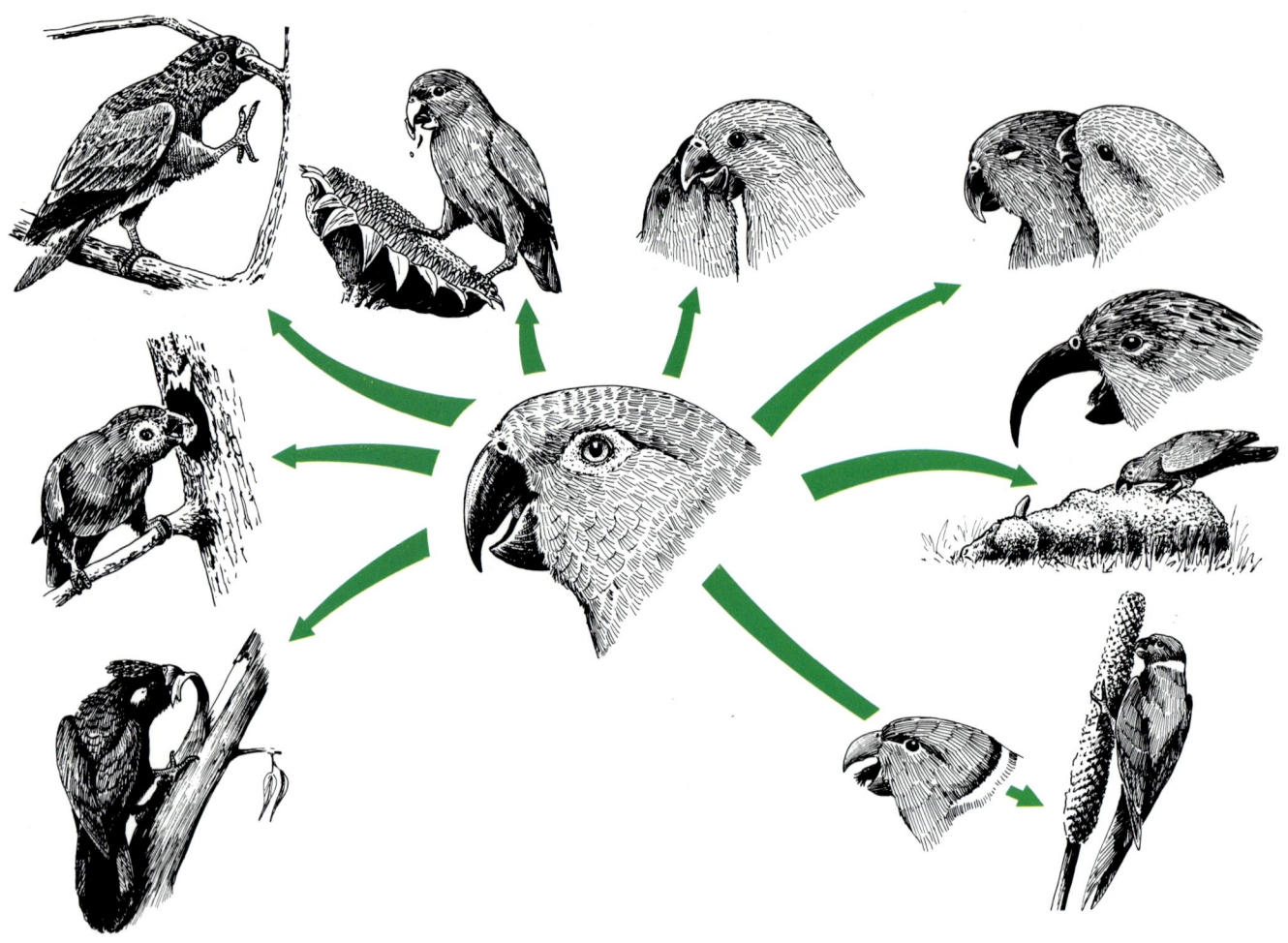

The characteristically hooked bill of the various members of the parrot family is used for a variety of purposes – stripping bark, drilling holes, grasping branches and so forth. It is important for such social activities as mutual grooming and conjugal feeding; and it is, of course, the principal means of obtaining food, whatever the diet. Both the bill (as in the carrion-eating kea) and the tongue (as in the lorikeets) may be structurally modified for feeding purposes.

simple and betraying the obvious relationship between the tiny lovebirds and budgerigars and the massive cockatoos and macaws.

The highly specialised characteristics of the Psittaciformes have made it extremely difficult for ornithologists to discover obvious links with any other groups of birds. Such relationships as have been suggested are hypothetical rather than factual. Suggestions that they may, for example, be related to the Strigiformes have to be treated with great caution and it would seem possible that they have closer links with the Cuculiformes, particularly cuckoos (Cuculidae) and touracos (Musophagidae). But even such a relationship must be sought far back in time for fossil remains of parrots date from the Tertiary period, indicating that even 60 or 70 million years ago the Psittaciformes had already developed their own distinctive features.

The order comprises a single family (Psittacidae) which is tentatively divided into six subfamilies, 82 genera and 317 species. No other group of birds displays such a degree of anatomical uniformity. Apart from a few types which are regarded as aberrant, the homogeneity of basic form is such that division into separate subfamilies is somewhat arbitrary. The bill is invariably short, thick, downward-curving and strongly hooked, with a fleshy cere at the base in which the nostrils are set, this sometimes being wholly or partially covered with feathers. The short neck and compact body give the birds a heavy, robust appearance. The wings, usually of moderate length, are rounded and supported by powerful pectoral muscles, suitable (in the majority of species) for rapid and involved aerial manoeuvres. Very few of the members of the family, however, are proficient at gliding and are normally compelled to make frequent halts for rest.

The dimensions of the tail are variable. In some of the smallest parrots it is almost invisible, hardly projecting beyond the tips of the wings, but in others, such as the macaws, parrakeets and budgerigars, it is fairly long in relation to the size of the body.

The short tarsi terminate in four toes, two in front and two behind, covered with large scales, enabling the birds to maintain a firm grasp on branches, to move confidently about in trees and to perform all manner of acrobatic manoeuvres.

The bill too is invaluable in providing an additional point of support, hooked into bark and around branches. The upper mandible, incidentally, is not firmly fused to the skull but is hinged to it, thus permitting much freer movement than is possible with other birds.

Bill and claws are also used in conjunction when feeding. Apart from certain carnivorous species such as the New Zealand kea, which has acquired the reputation of attacking live sheep in order to consume their fat, parrots are essentially vegetarians, feeding principally on seeds and fruit. The inner edge of the tip of the upper mandible is grooved in such a way that the bird can retain a grip on the toughest seeds whilst removing the husks

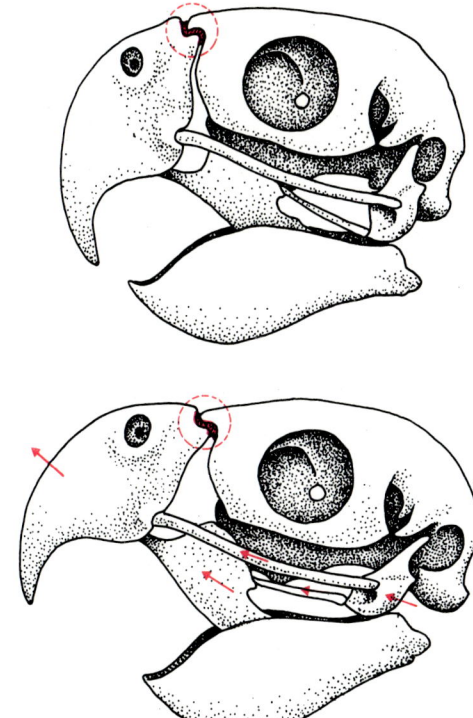

The upper mandible of a parrot is hinged to the skull, permitting a greater degree of mobility than is the case with most other birds.

Cockatiel
(*Leptolophus hollandicus*)

Red-sided parrot (male)
(*Lorius pectoralis*)

Red-sided parrot (female)
(*Lorius pectoralis*)

Turquoise parrot
(*Neophema pulchella*)

Bourke parrakeet
(*Neophema bourkii*)

Yellow-tailed black cockatoo
(*Calyptorhynchus funereus*)

Crimson rosella
(*Platycercus elegans*)

The rosellas usually have short, rounded wings and a grey or black bill. The colour of the plumage is very variable as can be seen here in photographs of the pale-headed rosella (*left*) and the eastern rosella (*right*).

much more effectively than if the two sections of the bill were interlocking. If by chance a fragment of food should be too cumbersome to be held in this manner the bird will grasp it between its toes and raise it to beak level, keeping its balance with the other foot. Either way the bill is powerful enough to break open the morsel concerned.

The tongue is rounded, fleshy and extremely mobile, sometimes fringed and terminated by bristles—a most effective organ for extracting food enveloped in a husk or shell.

With few exceptions male and female Psittacidae do not differ much in size, but often show differences in plumage. The majority are forest dwellers, congregating in noisy, chattering flocks, but in Australia a number of species are commonly found in sparse woodland and open country. These will spend the greater part of the time on the ground searching for food, only ascending trees for nesting purposes. Quite frequently they are to be seen

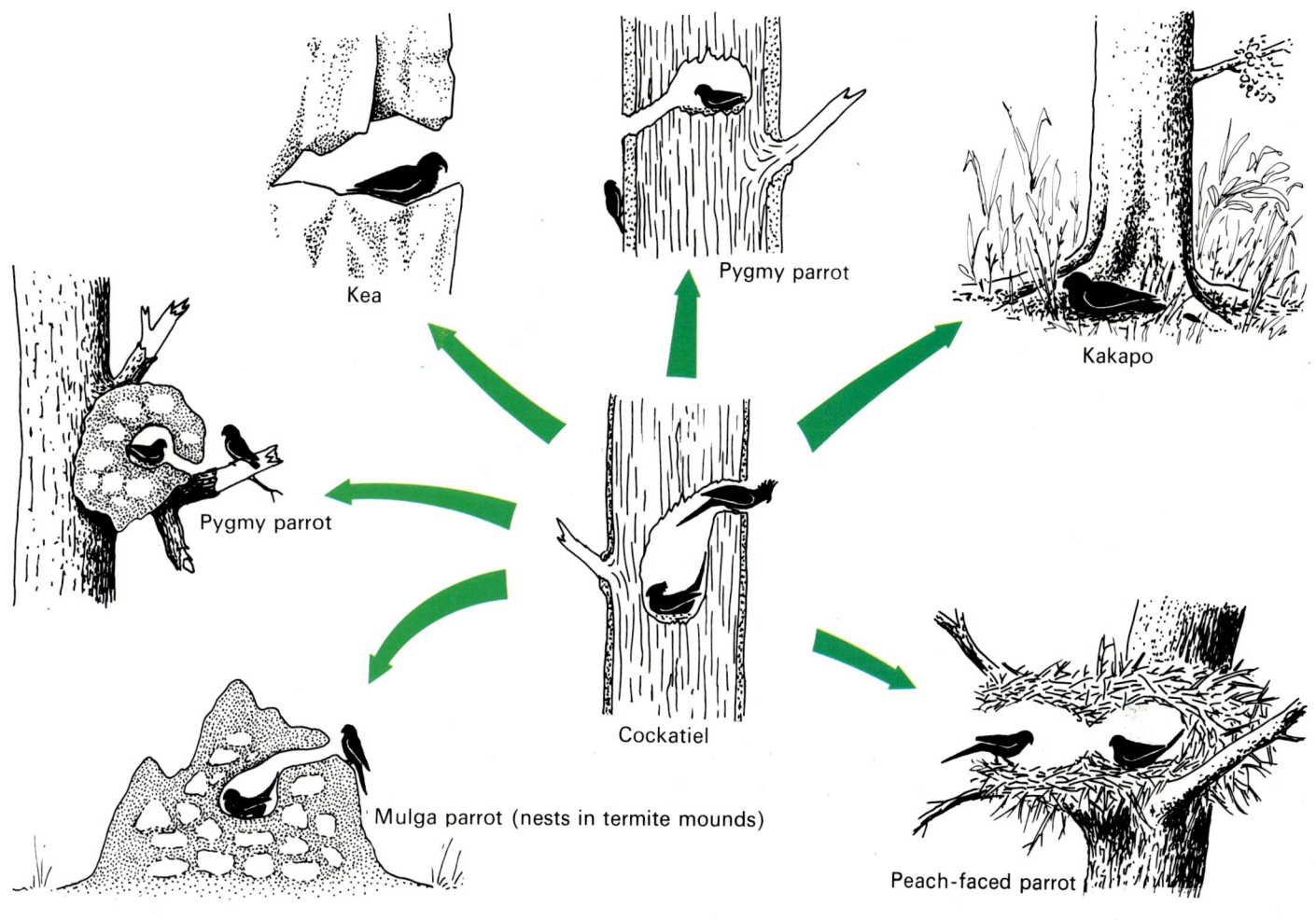

eating cultivated crops and because of their numbers and appetite are capable of causing extensive damage to farmland in some areas.

Most of the Psittaciformes build their nests in tree hollows some distance from the ground. A few rare species, however, nest in rock openings or scoop tunnels in cliffs or embankments. The nest itself is simple in structure and usually the eggs are laid inside without any prior preparation. But certain species assemble miscellaneous scraps of vegetation and other substances to line their nest. Thus the grey-headed or lavender-headed lovebird, or whitehead (*Agapornis cana*), a Madagascan species, has the strange habit of embellishing its nest with wood shavings, vegetable fibres and leaves which are carried tucked into the feathers of the back. In almost every species nests are constructed some distance apart; only one South American parrot nests in colonies.

Between two and five eggs are normally laid and these are pure white, characteristic of birds habitually nesting in cavities. After about three weeks' incubation the young are born blind, naked and defenceless, being fed by their parents on predigested food.

Parrots in the wild are extremely noisy, with a wide range of squawks and shrieks; but the ability of some species to mimic human speech and other sounds seems to be confined to birds in captivity. It is this entertainment value as well as their colourful appearance which makes them such popular cage birds.

Although most parrots build a nest in the rotten wood of a tree, the actual site, shape and constituent materials vary considerably according to species, as shown in this diagram.

| AUSTRALASIAN COCKATOOS |
|---|
| Class: Aves<br>Order: Psittaciformes<br>Family: Psittacidae |
| **SULPHUR-CRESTED COCKATOO**<br>(*Kakatoe galerita*) |
| Total length: 20 inches (50 cm)<br>Diet: fruit, shoots, leaves, seeds, etc<br>Number of eggs: 2–3<br>Incubation: about 30 days |
| Ivory-white plumage with yellow crest, cheek patch and undersides of tail feathers. Eye and bill black. Claws greyish. |
| **PALM COCKATOO**<br>(*Probosciger aterrimus*) |
| Total length: up to about 32 inches (80 cm)<br>Diet: fruit (especially of palm tree), leaves, shoots, seeds<br>Number of eggs: one<br>Incubation: about 30 days |
| Largest of cockatoos, with blue-black plumage and black crest. Naked pink or red patches on cheeks. Bill of male very strong, larger than that of female. |
| **COCKATIEL**<br>(*Leptolophus hollandicus*) |
| Total length: 12 inches (30 cm)<br>Diet: vegetation<br>Number of eggs: 4-7 |
| Grey plumage, white wings, conspicuous yellow crest, yellow cheeks with pinkish-orange patch; fairly long tail. Female and young less brilliant than male. |

*Facing page:* Cockatoos are distinguished from other Psittaciformes by a crest of feathers on the head which can be raised or lowered at will. In the sulphur-crested cockatoo the yellow crest stands out in contrast to the white plumage.

# The noisy, gregarious cockatoos

Australia has many wonderful surprises in store for animal lovers but one of the most remarkable and exhilarating sights is surely that of a flock of cockatoos, circling in an immense white cloud above plains and forests and swooping down in a chattering throng to land among eucalyptus trees, in a park or an open field, or on the shores of a lake.

The species in question is the sulphur-crested cockatoo (*Kakatoe galerita*), probably the best known of all local species and frequently seen in zoos abroad. Inspected at close quarters the individual bird is equally striking. In contrast to other parrots but in common with other members of the subfamily Kakatoinae, each bird bears on its head a magnificent crest of long yellow feathers which can be erected or lowered at will. Apart from yellow on the cheeks and tail feathers the only contrasts to the snowy white plumage of the rest of the body are the black eye and the typically hooked black bill.

Although colours vary (the largest species is black with red cheeks) cockatoos have common anatomical features and the subfamily is divided into five genera, seventeen species and some eighty-five subspecies. All are fairly large birds, most of them measuring between 15 and 20 inches but some exceeding 30 inches in total length.

The bill of a cockatoo is thick and heavy, in certain cases as large as the rest of the head. The upper mandible is much more massive than the lower and the inner edges of both are deeply grooved. The cheeks may either be naked (as in the palm cockatoo) or covered with feathers. The characteristic head crest is usually of a contrasting colour to the rest of the plumage and is probably used for purposes of intraspecific communication.

It is interesting to note that cockatoos, like herons, produce a powdery substance used for cleaning the plumage. Furthermore there is a well developed uropygial gland at the root of the tail exuding an oily secretion which is similarly used for preening the feathers.

The female cockatoo normally lays from two to four eggs (more in some species) and both partners incubate them for 21–30 days. The young leave the nest and make their first attempts to fly about two months after hatching.

The aforementioned sulphur-crested cockatoo is intelligent and easily tamed. In captivity it is one of the best 'talkers' of the parrot world. Its range of distribution covers all parts of Australia except for the central desert regions and the dry eastern coastal belt, New Guinea, the Indonesian islands of Aru and Seram, and the islands of the Bismarck archipelago.

In the wild these birds are highly gregarious, assembling in huge flocks save in the breeding season. They tend to remain as close as possible to water and are most frequently seen perching in tall trees bordering lakes and rivers, although they may fly some distance in quest of food.

At dusk each flock returns to its habitual roost and the deafening cacophony as the birds dispute places of vantage may

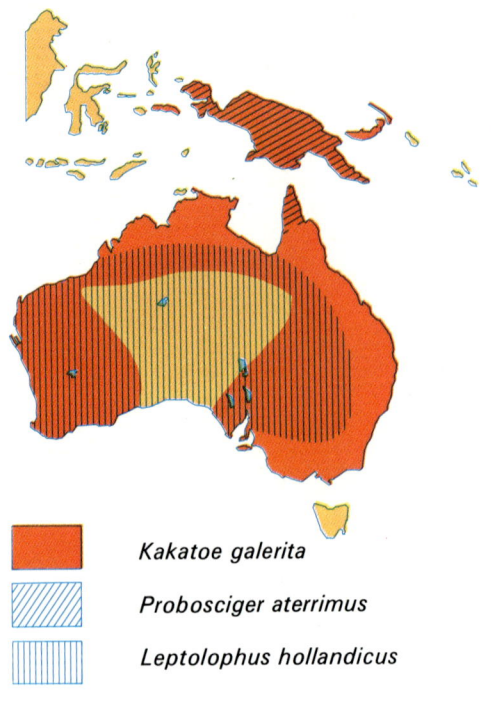

Geographical distribution of the sulphur-crested cockatoo (*Kakatoe galerita*), the palm cockatoo (*Probosciger aterrimus*) and the cockatiel (*Leptolophus hollandicus*)

continue until it is dark. As soon as dawn breaks the cockatoos are up and away, paying an initial visit to the nearest source of fresh water, this also being the last port of call before retiring for the night. Much of the ensuing day is spent roaming about for food, principally seeds of herbaceous plants, supplemented by insects and, if available, fruit, berries and nuts. Cereal crops are consumed in large quantities in some areas but recent investigations suggest that the damage done by the birds in this respect is more or less balanced by their useful consumption of grass seeds. During the hottest hours of the day the birds seek the shade of trees.

The courtship ritual of the sulphur-crested cockatoo is comparatively uncomplicated. The male struts back and forth ostentatiously, then turns towards his intended partner with crest erect, lifting and dipping his head or jerking it from side to side, all to the accompaniment of soft calls. Once a pair is formed each bird settles down to preen its plumage or to groom its partner.

Although pairs of sulphur-crested cockatoos may be seen at virtually any time of the year, sexual activity seems to be at its peak between August and January in the southern parts of Australia and between May and September in northern areas.

The female lays two or three eggs in the hollow of a tree (generally a eucalyptus), although occasionally she will deposit them in a rock cavity. Incubation lasts approximately one month and some fifty days may elapse before the youngsters are

ready to embark on their first flights. It is noticeable that the adults are very cautious and silent when attending to their brood and that only when they venture some distance from the nest do they begin to engage in their customary noisy chatter.

## The spectacular palm cockatoo

The largest member of the cockatoo tribe and certainly the most striking in appearance is the palm cockatoo (*Probosciger aterrimus*). Measuring up to 32 inches in total length, this bird has blue-black plumage, the crest of long, pointed feathers being the same colour, but there is contrast in the naked pink or red patch beneath each eye.

The bill of the palm cockatoo is exceptionally large and is used both as a tool for cracking nuts and seeds and as a formidable weapon. A captive palm cockatoo which is not provided with sufficient space will often use this powerful bill to snap the bars of its cage or to shatter plates and bowls used for food and water. As in the case of the sulphur-crested cockatoo the two sexes differ slightly in appearance, chiefly with regard to the size of the bill, that of the female being smaller and consequently weaker than that of the male. In the young the feathers of the underparts are fringed with yellow.

There is no mistaking this cockatoo in the wild for its size and colour distinguish it from all other members of the subfamily.

The cockatoos are classified in a separate subfamily–the Kakatoinae. They are medium-sized or large birds, mostly gregarious by habit, and although the most familiar species are white some are black, as, for example, the red-tailed black cockatoo (*left*) and gang-gang cockatoo (*centre*), with its contrasting small red crest. The cockatiel (*right*) is a much smaller type of parrot, found throughout Australia with the exception of coastal regions.

Although in comparison with that of the sulphur-crested cockatoo its range is restricted, it is still fairly abundant. In some regions, however, which are gradually succumbing to the advance of human settlement, and notably on the Cape York peninsula, there are fears for its future.

Palm cockatoos normally live in pairs or in groups of five to eight, being considerably less noisy and gregarious than other species. Although their characteristic habitat is the tropical rain forest they have lately begun to breed in other biomes, such as the tree-covered savannah. They are often seen, for example, in thickets of eucalyptus and especially in trees bordering rivers.

The female lays a single egg in a deep hole pecked into a tree trunk, often high above ground level. The breeding season evidently lasts from August until February and the young bird will not quit the nest before it is two months old. Truth to tell, however, very little is known about the breeding behaviour of this spectacular species.

Food consists in the main of flower-buds of palm trees and other succulent plants, as well as nuts, fruit, seeds and leaves. Analysis of stomach contents have also revealed the presence of numerous pieces of quartz which apparently have the effect of assisting the digestive processes.

The palm cockatoo, unlike most of its relatives, is not only an excellent flier but also an accomplished glider. In the course

of such glides the wings are held motionless, pointing towards the ground.

The naturalist J. M. Forshaw, who has studied the habits of the species in some detail, reports that each bird sleeps in the higher branches of a tree on the borders of the rain forest, well apart from its companions, and that it does not leave its roost until the sun is high in the sky. But although a typical forest resident, it will often fly considerable distances over open ground in search of food, returning to its tree roost in the afternoon. The same ornithologist has also described the courtship rituals of these cockatoos, in which the males abandon their normal perching posture, bending forwards, stretching the neck, erecting the long feathers of the crest, spreading the wings and holding the tail upright. Recent observations suggest that such body movements may be characteristic of all individuals, both adults and young.

## Clouds of colour

Other cockatoos emulate the sulphur-crested species in forming immense flocks. One such is the little corella (*Kakatoe sanguinea*); another is the galah (*Kakatoe roseicapilla*) which is found throughout Australia except in the eastern and western coastal belts and is a familiar resident of city parks.

The gang-gang cockatoo (*Callocephalon fimbriatum*), with grey

Most cockatoo species roost in enormous flocks, setting out at daybreak in search of water and food. The galahs (*left*) have a wide distribution, being absent only from the narrow coastal belts of eastern and western Australia. Little corellas (*right*) are found in central Australia and are compelled to fly considerable distances to quench their thirst.

*Facing page*: The tongue of a parrot is fleshy (1) but in the lorikeets it is fringed with bristles, serving to extract nectar and pollen from flowers (2). The inner edge of the upper mandible is grooved (3), enabling the bird to grasp seeds and nuts firmly and to retain a grip on branches.

There are no true woodpeckers (2) in the Australasian region and the same ecological niche is here occupied by certain members of the parrot family, notably the yellow-tailed black cockatoo (3) and the various species of pygmy parrots (1). The Australasian birds clamber up and down vertical trunks, with the tail as an additional point of support, using their bill to drill into bark or strip it off in layers to feed on insects and larvae in much the same fashion as woodpeckers.

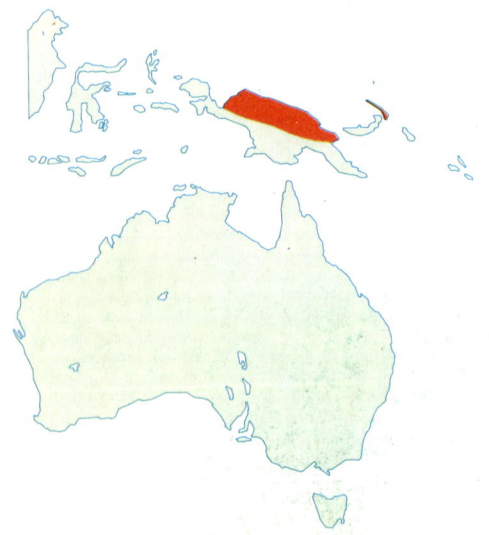

Geographical distribution of Bruijn's pygmy parrot.

plumage and a contrasting red head, is a much rarer species, confined to south-eastern Australia; and the delicate cockatiel (*Leptolophus hollandicus*) is another gregarious species notable for its delightfully coloured plumage. The body and the long tail are pale grey, the wingtips white, the front of the head and the erectile crest pale yellow, the cheeks pinkish-orange. Large flocks of these birds may be seen almost everywhere, apart from the northern and eastern coastal belts and the areas south of Perth.

Finally mention must be made of the yellow-tailed black cockatoo (*Calyptorhynchus funereus*), which is especially remarkable for its adaptive capacities. With the aid of its strong bill this bird systematically strips the bark off dead eucalyptus trees, piercing holes in the trunk that may be as large as a human hand in order to feed on xylophagous insect larvae. There are no woodpeckers in the Australasian region and this cockatoo occupies the same ecological niche and feeds on the same prey as the Piciformes in other parts of the world. In fact the yellow-tailed black cockatoo performs a valuable service by controlling local insect populations.

## Midgets of the parrot world

The six species of pygmy parrots which are inhabitants of New Guinea and adjacent islands make up a separate subfamily—the Micropsittinae. All the representatives of this group, belonging to the genus *Micropsitta*, are tiny (the smallest members, in fact, of the parrot family), ranging from 3 to 5 inches in length. They are beautifully coloured, generally with green back and flanks, dark wings and yellow or red underparts, sometimes fringed with cobalt blue. There is some sexual dimorphism, the females being less gaudy than the males.

These pygmy parrots are usually to be found in low-lying tropical rain forests although there is one species that lives in the mountains. All are strictly arboreal.

The tarsi and toes of these tiny birds are long in comparison with those of other parrots. But the really remarkable feature is the tail, the rectrices being short and robust, the tips of the rachis lacking barbs and consequently trimmed, facilitating their characteristic activities. For these pygmy parrots look and behave like woodpeckers, gripping the bark firmly with their claws, flattening the body against the vertical trunk and using the tail as an extra point of support.

---

**RED-HEADED OR BRUIJN'S PYGMY PARROT**
(*Micropsitta bruijnii*)

Class: Aves
Order: Psittaciformes
Family: Psittacidae
Total length: 4 inches (10 cm)
Weight: about ½ ounce (28 g)
Diet: sap of certain trees, termites, fruit
Number of eggs: probably 2

Together with five related species, the smallest member of parrot family. Reddish head and cheeks, dark green back, pinkish-yellow underparts, blue tail feathers. Relatively long feet and toes. Bluish-violet band extends across flanks and crop and around neck. Tip of short tail lacks barbs.

By using the tail to secure their foothold the pygmy parrots are capable of performing acrobatic feats that are even more spectacular than those of the woodpeckers they so closely resemble, scampering up and down, sideways or at an angle with astonishing speed. Like the yellow-tailed black cockatoo these little birds seem to have copied the woodpeckers in almost every detail of behaviour, yet another fascinating example of convergent evolution. But apart from this it must be admitted that precious little is known about their biology. Examination of stomach contents shows that the remains of food, with the exception of such definable items as seeds and termites, are made up of a white pasty substance of completely unknown origin. Ignorance of the principal constituents of their diet is the main reason why it has not been possible to keep these birds in captivity, for the few that have been caught for purposes of scientific study have not survived very long. According to local tribes (whose opinions can never be lightly dismissed) the birds feed chiefly on insects. Recent investigations have convinced many ornithologists, however, that they also eat certain types of fungi, this view being supported by the fact that they are often seen clinging to the decaying bark of dead trees. This has led naturalists to the conclusion that the mysterious pasty substance is of fungal origin, lack of which causes the death of zoo birds. Yet feeding captive pygmy parrots with fungi has not had the desired effect for the birds have continued to die, suggesting to some experts that the real cause of death may be

The blue-crowned pygmy parrakeet is one of six species from New Guinea, smallest members of the parrot family, which look and behave much like woodpeckers. The long tail, the tip of which lacks barbs, helps to support the bird's body when balancing on tree trunks.

stress brought about by loss of liberty—enough to upset body metabolism. They would not be the first animals to languish when deprived of their freedom.

Breeding habits of pygmy parrots appear to differ little from those of other Psittaciformes, with eggs being laid in the hollow of a dead tree or in the nest of tree-dwelling termites. Such a nest will usually be situated about 10-15 feet above the ground, with an entrance just wide enough to encompass the body of one of the parents. The tunnel is drilled in an upward direction so that the incubation chamber is at a slightly higher level than the entry hole. This is an unusual feature which causes few problems to birds that exhibit such superior climbing skills.

In the event of a pygmy parrot taking up residence in a termitarium, the insect occupants get busily to work blocking all the tunnels leading to the intruder's nest, with the result that the birds are undisturbed by the termites at any stage of breeding activity.

The eggs of all species of pygmy parrot are white and no more than two nestlings have been positively identified. Some authors believe that the male excavates a separate hole for sleeping but surveys in one area have shown that the two adults normally occupy the same hollow. In some nests up to six birds were located but ornithologists were unable to determine whether these were adults or young of previous broods.

## The rainbow-hued lorikeets

Some of the most beautiful members of the parrot family are the Trichoglossinae, a large and distinctive subfamily consisting of 61 species and 150 subspecies made up of lories and lorikeets— in Australia they are usually all called lorikeets—very similar in appearance to one another.

The most common representative of the subfamily Trichoglossinae is the rainbow lorikeet (*above and below*) whose plumage is a magnificent combination of vividly contrasting colours, predominantly red, yellow and blue.

*Facing page:* The various lorikeets feed on the juice and pollen of flowers and are often seen perching acrobatically on the thinnest branches of flowering trees—like the little lorikeet (*Glossopsitta pusilla*) shown here.

Geographical distribution of the rainbow lorikeet.

**RAINBOW LORIKEET**
(*Trichoglossus mollucanus*)

Class: Aves
Order: Psittaciformes
Family: Psittacidae
Total length: 7½–11 inches (19–28 cm)
Diet: nectar and pollen, fruit, shoots, seeds and insects
Number of eggs: 2–4
Incubation: 21–26 days

Beautiful bird with blue head, red bill and iris, green back, pale green band on neck, blue belly, orange breast, crimson lower tail coverts, black remiges flecked with white.

Like most parrots rainbow lorikeets, which form enormous flocks, are gregarious birds ranging from the east coast of Australia to New Guinea and adjacent islands, including the Celebes. At least twenty subspecies have been identified.

This group, whose most familiar representative is probably the rainbow lorikeet, comprises birds measuring anything from 5 to 14 inches (thus ranging in size from that of a sparrow to that of a large turtle-dove). The birds are distributed fairly widely through the Australasian region, their range extending from Australia to New Guinea and the Polynesian islands. All are brightly coloured, the larger species having a long tail, the smaller ones possessing much shorter rectrices. The majority are gregarious, tending to form noisy flocks. The bill is short and curved and the tongue has a brush-like appearance, similar to that of the honeyeaters. This tongue with its covering of soft bristles is useful for sipping nectar and pollen from flowers, above all those of *Eucalyptus* species. Because of their distinctive feeding habits these birds play an important role in plant pollination with a consequent effect on tree growth and forest development. The birds also consume fruit and shoots.

Lories and lorikeets are easily tamed and are familiar and popular residents of zoos and aviaries.

The large members of the subfamily lay two eggs and rear their young for about fifty-five days. Smaller species usually lay three or four eggs and their nestlings take rather less time to assert their independence.

The well known rainbow lorikeet (*Trichoglossus mollucanus*) inhabits the whole eastern seaboard and Tasmania, while the similar red-collared lorikeet (*Trichoglossus rubritorquis*) is along the north-west coasts of Australia, the most northerly parts of Dampier Land and Arnhem Land, Celebes, Timor, the Lesser Sunda Islands, New Guinea and New Caledonia. With its green back, red bill, orange breast and crop, and cobalt blue head, the rainbow lorikeet is indisputably one of the most beautiful of all parrots. Extremely gregarious, the birds form immense chattering flocks that roam far and wide, frequently in the company of other species. Flowering trees, especially eucalyptus, attract hordes of these lorikeets for the simple reason that the corollas provide them with their main sustenance in the form of nectar and insects. At such seasons the birds make a spectacular picture as they cling in all kinds of acrobatic poses to the branches, a sight not necessarily appreciated by farmers who are more concerned to protect their blossoming fruit trees.

Like the majority of Psittacidae the rainbow lorikeets nest in the hollows of tree trunks. The female alone apparently incubates her eggs, the male visiting her at dawn and dusk. As one might expect from their seasonal wanderings over forests and orchards, the birds are excellent fliers; and despite their brilliant colours they can pass almost unobserved in the flickering lights and shadows of the woodland scene.

The species is of great scientific interest because it exhibits, on a small scale, the phenomenon of evolution and diversification. Throughout their range of distribution, which although enormous is not continuous, the birds have taken up residence in regions that are separated from one another by ocean or desert, enabling ornithologists to distinguish at least twenty different races. In fact the rainbow lorikeet has a claim to be considered a super-species, the varieties stemming from it possessing

Feeding

Climbing

Parrots have four toes on each foot, two of which are directed forwards and two backwards. By this means they can use one foot to lift a morsel of food to the mouth, cracking the husk of a seed, for example, with the bill and are able to maintain a firm grasp with the toes whilst climbing up, down or along a branch.

Geographical distribution of Pesquet's parrot.

---

**PESQUET'S PARROT**
(*Psittrichas fulgidus*)

Class: Aves
Order: Psittaciformes
Family: Psittacidae
Total length: 20 inches (50 cm)
Diet: vegetation

Strange-looking large parrot, the front of the head naked except for a few sparse, coarse feathers. Strong, slightly curved bill. Colour greyish-black with bright red wings and belly.

---

*Facing page:* The Australasian region is the home of many delightful little parrakeets which have adapted to their surroundings in different ways. The popular budgerigar (*left*) inhabits arid regions and nests in tree hollows. The ground parrot (*below right*) has a mottled plumage which blends well with the undergrowth in which it nests. The blue-winged parrakeet (*above right*), like others of the genus *Neophema* is also a ground bird which is often seen running rapidly through the grass.

enough differentiating features to justify their being raised to species status.

Closely related to it are three other species of the same genus, the varied lorikeet (*Psitteuteles versicolor*), the ornate or ornamented lorikeet (*Trichoglossus ornatus*) and the scaly-breasted or gold and green lorikeet (*Trichoglossus chlorolepidotus*).

## Land of the budgerigars

There is no more popular cage bird today than that small member of the parrot family, the budgerigar (*Melopsittacus undulatus*). But although its behaviour behind bars is familiar enough, a simple matter of day-to-day observation, the owner of such a pet probably knows little or nothing about the background of this species and its totally different behaviour in the wild.

This sociable, garrulous, cheerful bird originates in Australia where it enjoys an immense area of distribution covering the driest regions of the interior but excluding the coastal woods and rain forests.

The first naturalist to bring this charming bird to the notice of Europeans was John Gould. In 1840 the famous British ornithologist introduced several specimens of the species to England and it was not long before the birds were being imported in their thousands, initially to the British Isles and subsequently to other European countries.

The Aborigines of the Sydney region gave this bird the name of *betcherrygah* which signified 'good food' and which at the same time roughly reproduced its call. From this was derived the name budgerigar. Gould himself described it as a singing grass parrakeet and the first breeders labelled it canary parakeet. As frequently happens when a species is bred in captivity in large numbers, a multitude of colour forms have since been produced so that nowadays there is a wide choice—white, yellow, green, blue and mauve—in a variety of subtle and delicate shades and patterns. But in the wild the dominant colour is green or greenish, the wings and back being streaked with brownish-black bars. The head is yellow with a cluster of thin black bars and the long tail is blue. The comparatively small hooked bill is almost invisible beneath the forward-jutting feathers of the cheeks.

Tail apart, the budgerigar is slightly smaller than a sparrow. There is a slight degree of sexual dimorphism for the male, in common with other parrots, has a stronger beak and a more swollen cere, the latter being violet-blue. In the female the cere is pale blue but darkens to become brownish while she is rearing the young.

According to Forshaw, one of the leading authorities on the parrot family, the budgerigar belongs to a separate genus with only one species and appears to be a transitional form between the little grass parrots of the genus *Neophema* and the flightless ground parrot of the genus *Pezoporus*.

Budgerigars are inhabitants of deserts and semi-desert regions where they congregate in huge flocks that may number tens or even hundreds of thousands and which literally blacken

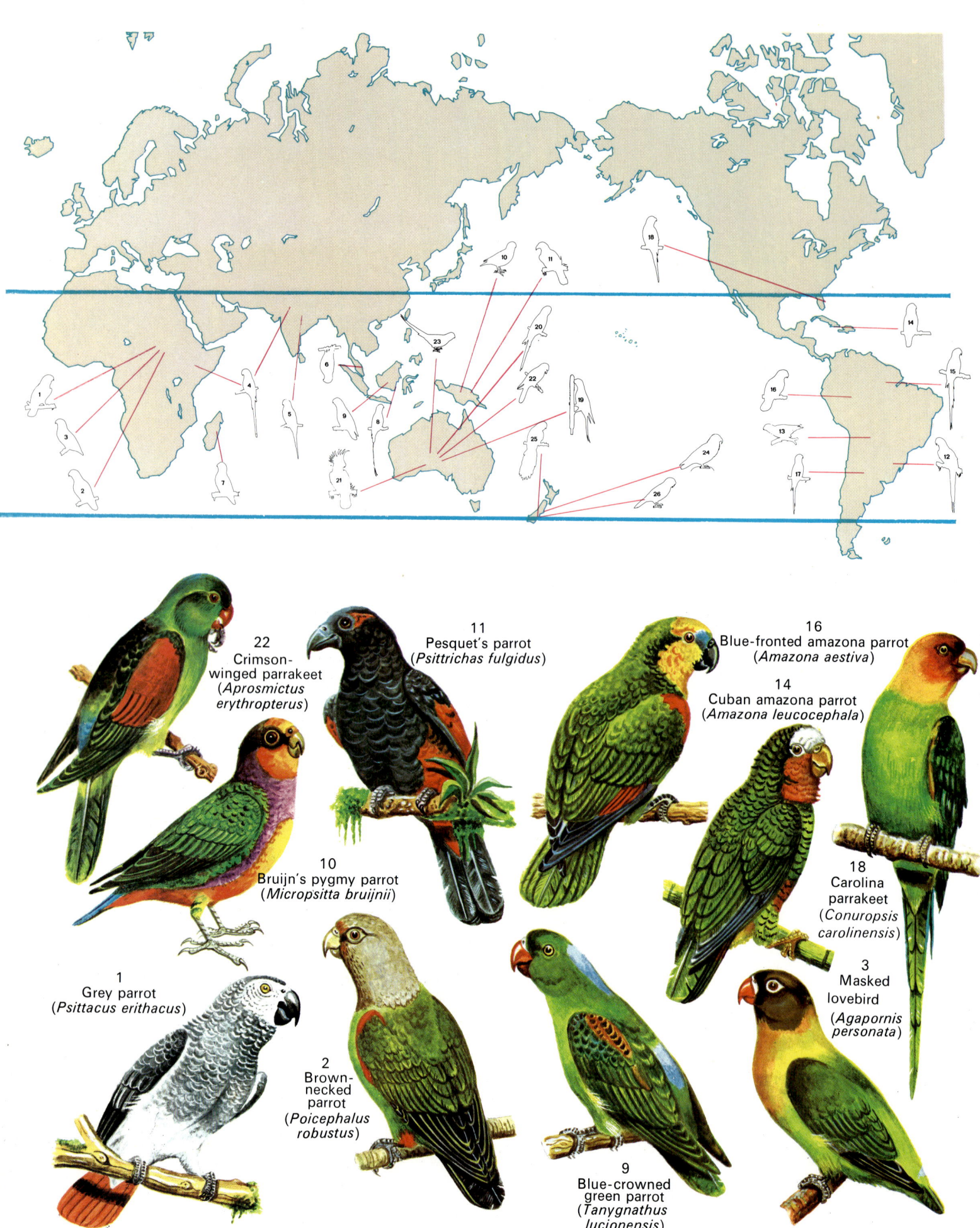

# Order: Psittaciformes

The parrots, parrakeets, cockatoos, lories, lorikeets, lovebirds, budgerigars and macaws that are grouped together in the order Psittaciformes all belong to a single family, Psittacidae. The members of this family are basically similar to one another, possessing characteristic features which permit immediate identification and which clearly differentiate them from other families and orders of birds. Thus the bill is massive and strongly hooked, the upper mandible being larger than the lower. The inner edge of the upper mandible is grooved, which helps the bird to take a firm grip on seeds when feeding and to use the bill for grasping branches, thereby providing tree-dwelling species (the vast majority) with an additional support to the feet. There are four toes on each foot, two directed forwards, two to the rear.

With rare exceptions, such as the kakapo of New Zealand and the Australian ground parrot, neither of which is capable of flying, the Psittacidae are powerful and rapid fliers although few are accomplished gliders.

Food consists fundamentally of fruit, leaves, shoots and seeds but some species feed on small animals, such as the yellow-tailed black cockatoo of Australia which consumes the larvae of wood-boring insects, and the pygmy parrots of New Guinea which apparently eat termites. The tongue, fleshy and rounded, helps most parrots to extract seeds; in the Trichoglossinae, however, it is covered with bristles and thus assists the bird in sipping nectar and pollen from flowers. In general all parrots nest in tree hollows, although some choose rock fissures.

The order comprises 79 genera and 317 species which are usually grouped in the following subfamilies:

Nestorinae. Two species of fairly large New Zealand parrots belonging to the genus *Nestor*. The kea (*Nestor notabilis*) is a highland bird nesting in rock clefts which turns scavenger in winter to supplement its regular diet of vegetation, insects and invertebrates, sometimes attacking domestic sheep to peck at their fat. Closely related is the kaka (*Nestor meridionalis*), a bird of lowland forests.

Strigopinae. The only species is the kakapo or owl parrot (*Strigops habroptilus*), also from New Zealand and nowadays threatened with extinction. It is a nocturnal ground bird which has lost all power of flight but is capable of climbing trees and gliding to the ground.

Trichoglossinae. Lories and lorikeets, small and medium-sized birds with brush-like tongue, enabling them to lap juices from flowers that are crushed with the beak. This characteristic is not found in the genera *Psittaculirostris* and *Opopsitta*.

Micropsittinae. Pygmy parrots, the smallest representatives of the parrot family, inhabitants of the New Guinea rain forests, all belonging to the genus *Micropsitta*. All are tree-dwellers, climbing and pecking rotten bark in the manner of woodpeckers. Food reputed to consist of termites and other insects and possibly including certain fungi.

Psittricasinae. One species, Pesquet's parrot (*Psittrichas fulgidus*), from the northern part of New Guinea. Vegetarian diet, but little known of habits in the wild.

Kakatoeinae. Cockatoos, exclusive to the Australasian region. Large or medium-sized birds with distinctive head crest of long, pointed feathers. Familiar species, often seen in zoos, are the sulphur-crested cockatoo (*Kakatoe galerita*) and the pink cockatoo (*Kakatoe leadbeateri*). Largest species is the palm cockatoo (*Probosciger aterrimus*).

Psittacinae. Large subfamily distributed through South America, Africa and the Oriental and Australasian regions, including parrots, parrakeets, lovebirds, macaws, amazons, rosellas and budgerigars.

The pattern or structure of the head feathers of many parrots evidently serves as a signal for others of the same species. In the pink cockatoo (1) it assumes the form of a crest; in the plum-headed parrakeet (2) a narrow black band around the throat stands out prominently; in the masked lovebird (3) the eye is surrounded by white 'spectacles'; and in the yellow-breasted macaw (4) there is a conspicuous bare white patch on the cheeks.

*Facing page:* The crimson rosellas found along the east and south coasts of Australia and in Tasmania show much colour variation according to the region inhabited. There is no evidence of cross-breeding between the different subspecies.

CHAPTER 101

# Rare species in peril: the flesh-eating marsupials

The settlement of Australia and the adjacent islands had tragic repercussions for indigenous forms of wildlife, following the introduction of foreign species. The marsupials of the Australasian region had evolved ever since the Cretaceous period in conditions of total isolation. Carnivores and herbivores alike found themselves unable to put up any significant defence against the placental mammals which arrived much later, the laws of natural selection applying here as everywhere else. Carnivorous marsupials were in fact worse affected than herbivores. Incapable of competing with the imported predators (foxes, cats, dogs) for food, and subjected to a vicious campaign of extermination on the grounds of harming domestic stock, some groups were completely wiped out, a grievous loss to science.

Zoologists believe that the first marsupials to colonise the Australasian region were small carnivores similar to certain surviving species such as the brush-tailed phascogale (*Phascogale penicillata*) and that these were the ancestors of all present-day marsupials. Seventy million years ago these animals lived in the northern hemisphere, and as their populations spread they eventually reached, by way of land links that have since vanished, the virgin territory of what is now the island continent of Australia, dispersing in all directions to occupy every available ecological niche. Older indigenous species–presumably monotremes or closely allied groups–were probably systematically supplanted, just as, at a later stage, these same marsupials were compelled to give way in the face of an invasion by the more highly evolved true mammals. Unfortunately we have no concrete record of this epic confrontation between marsupials and monotremes for palaeontologists have not thus far succeeded in discovering fossils which would undoubtedly have thrown

*Facing page:* The tiger cat or large-spotted native cat is the largest flesh-eating marsupial of the Australasian region. Resembling a mustelid more than a feline, it has the agility and intelligence of a typical tree-dwelling predator.

Examination of a few significant anatomical features is enough to provide reliable information about an animal's evolutionary development and life habits. This applies to marsupials as well as to placental mammals. Teeth and feet are especially revealing and on the basis of these, marsupials fall into two main categories. Some, like the kangaroo, possess dentition characterised by a single pair of incisors in the lower jaw (diprotodonts) and toes which are reduced in size or fused with one another, indications that the animal is a plant-eater and uses its legs for rapid escape from predators. Similar features are found in the horse, an equivalent placental mammal. Other marsupials, such as the tiger cat, have numerous sharp incisors in the lower jaw (polyprotodonts) with well developed, separate toes, furnished with powerful claws, indicative of a carnivorous diet and hunting habits. These characteristics are also present in the wild cats of the family Felidae.

The marsupium of the female bandicoot, in this drawing the long-nosed bandicoot, opens towards the rear, as is common with marsupials that move on all-fours.

light on the sequence of events. The principal reason for this is that Australia, being one of the few regions of the world not to have been subjected, over millions of years, to massive land upheavals or major climatic fluctuations, has not provided the geological conditions necessary for the preservation of fossils which would have enabled scientists to uncover many of the mysteries of life in bygone times.

## Evidence of tooth and claw

There are certain anatomical features of an animal which can provide more reliable information about its way of life than any generalised description. If a zoologist were asked to formulate the habits of a completely unknown species and given the choice of forming his deductions either from a photograph or from a single tooth of the animal in question, he would almost certainly select the latter. Although two animals with completely contrasted feeding habits may, as a result of adaptation to their surroundings, look superficially alike, the structure of their teeth will be quite different.

Thus in spite of the apparent diversity of form, identification and classification of marsupials is greatly simplified if one concentrates on a few well defined anatomical features. Examination of the jaws reveals two principal types of dentition, each with a certain number of secondary variations. Some marsupials have a tooth structure in the lower jaw which is characterised by a single pair of incisors and these are known as diprotodonts.

Others have a larger number of incisors (eight to ten) in the lower jaw, and these are described as polyprotodonts.

The structure of an animal's teeth can reveal much valuable information but the zoologist will, if possible, also examine the limbs, for the two features are usually closely linked. Thus the number of toes of diprotodont herbivores tends to be reduced, with one pair fused together, narrowing the surface in contact with the ground and thereby increasing the running capacity of an animal which is the natural prey of polyprotodont carnivores, the majority of which retain five toes. The study of the limbs is interesting for more than one reason. Not only may it reveal much about the behaviour of an animal but more detailed investigation may provide information about its environment. Long, sharp claws, for example, would be indicative of burrowing or tree-climbing habits and help to form a picture of its original surroundings.

Running is the life-saving device of a herbivore and consequently its limbs will be long and flexible, with the individual toes of minor importance, often undeveloped. In the case of carnivores, however, the problem is primarily one of strength rather than speed and good use is made of all five toes which are generally armed with sharp claws. Although they must be fast enough for the pursuit of prey over short distances the limbs

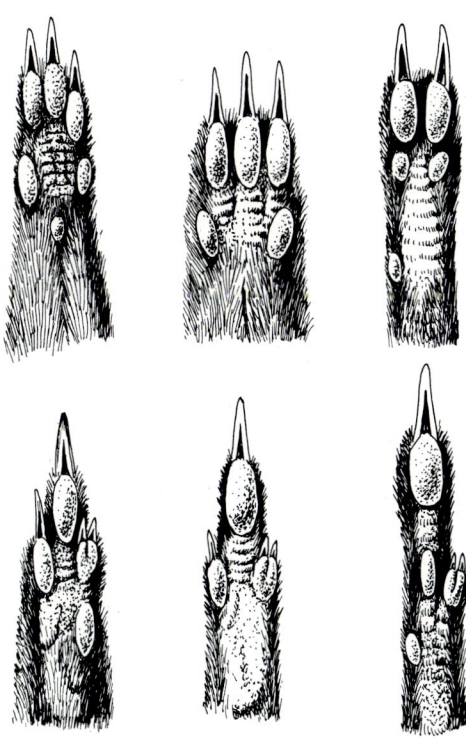

The bandicoots of the family Peramelidae are unusual in possessing the polyprotodont form of tooth structure common to carnivores combined with a toe structure characteristic of macropodid marsupials. This has come about as a consequence of their feeding habits which are either carnivorous or omnivorous. The toes of the various species are reduced to a lesser or greater degree and these drawings show the soles of the forefeet (*above*) and the hind feet (*below*) of three species of bandicoot. On the left are the feet of the short-nosed bandicoot which digs for worms and bulbs. In the centre are the feet of the rabbit bandicoot which excavates deep burrows, working with astonishing speed. On the right are the feet of the pig-footed bandicoot in which species the toes are reduced to a greater extent than in all others, indicating that the animal is an accomplished runner.

Bandicoots, which constitute the family Peramelidae, characterised by a long, pointed snout, have a mixed diet and are nocturnal.

The short-nosed bandicoot (*above*) is a solitary animal with strong territorial instincts. The rabbit bandicoot (*facing page*) can excavate a hole as fast as a man can dig.

Geographical distribution of bandicoots.

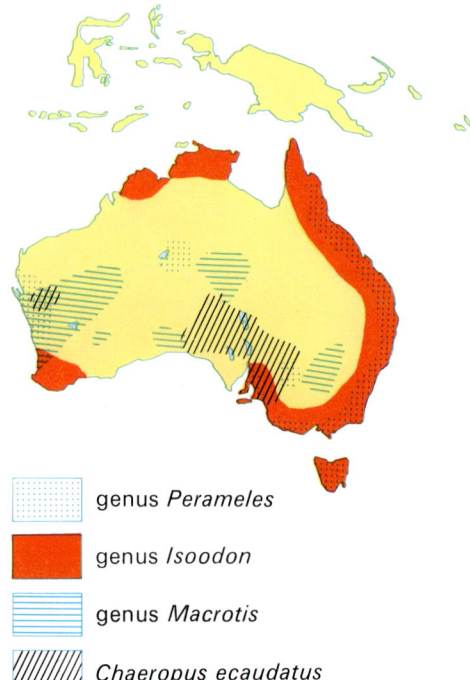

| | genus *Perameles* |
| | genus *Isoodon* |
| | genus *Macrotis* |
| | *Chaeropus ecaudatus* |

of carnivores must be suitable too for stunning, immobilising and killing. Hence the difference, for example, between the hoof of a horse and the paw of a tiger.

## The curious bandicoots

The strange marsupials known as bandicoots, belonging to the family Peramelidae, seem to contradict all that has just been written about the links between teeth and claws. For these animals have a polyprotodont tooth structure yet possess reduced, syndactylous toes that are typical of diprotodonts. Detailed study of their habits reveals the reason for this apparent discrepancy, the bandicoots being, in varying degrees, partially carnivorous and partially herbivorous. Having been subjected to opposing selective pressures, they cannot be classified either as specialised predators or as typical phytophages, with the result that their development has been a compromise in which the two types of anatomical feature have mingled.

One unmistakable characteristic of this group of marsupials is the long, slender snout, the extreme example of which occurs in the bandicoots of the genus *Perameles*. Another interesting feature is that they are one of the few genera of marsupials to possess the type of placenta found in other mammals. Even more remarkable, however, is the structure of the feet. The toes of the hind feet vary greatly in size, the second and third generally being much shortened, joined along their entire length and furnished with free claws whose function may be to clean the pelage. The first toe has been lost, while the fourth is enlarged and together with a smaller fifth toe provides the main support for the foot. The structure of the toes cannot be interpreted as

## BANDICOOTS

Class: Mammalia
Order: Marsupialia
Family: Peramelidae

### SHORT-NOSED BANDICOOT
(genus *Isoodon*)

Length of head and body: $9\frac{1}{2}$–$19\frac{1}{2}$ inches (24–49 cm)
Length of tail: $3\frac{1}{4}$–8 inches (8–20 cm)
Diet: omnivorous
Gestation: 2 weeks
Number of young: 2–5

Three species with short snout and small ears. Upper parts dark chestnut or yellowish-chestnut; underparts greyish-yellow, chestnut or white.

### LONG-NOSED BANDICOOT
(genus *Perameles*)

Length of head and body: 8–17 inches (20–43 cm)
Length of tail: 3–$6\frac{3}{4}$ inches (7.5–17 cm)
Diet: insects, vegetation
Gestation: about 2 weeks
Number of young: 1–5

Long snout, pointed ears. Upper parts pale chestnut with orange, greyish or yellowish tints; black hairs sometimes mingling with lighter ones.

### PIG-FOOTED BANDICOOT
(*Chaeropus ecaudatus*)

Length of head and body: 9–10 inches (23–25 cm)
Length of tail: 4–$5\frac{1}{2}$ inches (10–14 cm)
Diet: omnivorous
Number of young: usually 2

Large head and ears, pointed snout; long legs with much reduced toes. Upper parts greyish or orange-brown, underparts whitish. Possibly extinct.

### RABBIT BANDICOOT
(genus *Macrotis*)

Length of head and body: 8–20 inches (20–50 cm)
Length of tail: 4–8 inches (10–20 cm)
Diet: termites and insects, but will take flesh
Number of young: 1–3

Delicate, slender body, long pointed snout, very long ears. Soft, silky fur. Upper parts grey or blue-grey. Underparts white.

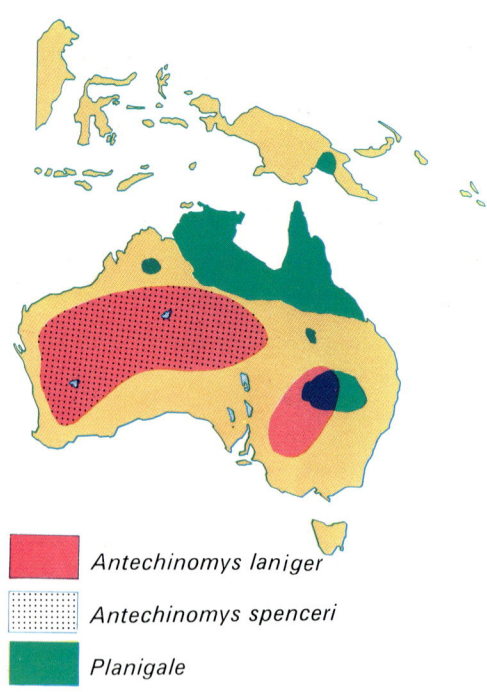

Geographical distribution of small dasyurids.

**JERBOA MARSUPIALS**
(*Antechinomys laniger* and *A. spenceri*)

Class: Mammalia
Order: Marsupialia
Family: Dasyuridae
Length of head and body: 3¼–4¼ inches (8–11 cm)
Length of tail: 4½–5¾ inches (11.5–14.5 cm)
Diet: insects, small vertebrates
Number of young: 6–8

Small head, pointed snout, large eyes and ears. Slender body, hind legs much longer than forelegs, long tail terminating in tuft. Upper parts greyish, underparts whitish.

**FLAT-HEADED MARSUPIAL MICE**
(genus *Planigale*)

Class: Mammalia
Order: Marsupialia
Family: Dasyuridae
Length of head and body: 1¾–3¾ inches (4.5–9.5 cm)
Length of tail: 2–3 inches (5–7.8 cm)
Diet: insects
Number of young: up to 12

Smallest known marsupials with extremely flat head. Thick pelage, tail covered with short hairs and without tuft. Upper parts greyish-red, underparts similar but lighter.

proof of any relationship with kangaroos but as an original and independent family feature. Bandicoots can stand upright on the hind legs, using the tail as a third point of support but cannot jump like kangaroos, and move about on all-fours. In keeping with this method of locomotion the female's marsupium opens to the rear.

Bandicoots are useful little animals in that their carnivorous diet consists basically of insects and their larvae and rodents. Unfortunately they are not officially protected and are often harassed for allegedly damaging fields and gardens in the course of their burrowing activities. Consequently some species are in danger of extinction.

The short-nosed bandicoots of the genus *Isoodon*, so called because the snout is not so well developed as that of other members of the family, are timid when faced by animal or human enemies but highly aggressive towards one another. Territory is clearly defined; the domain of each male, larger than that of a female, seldom overlaps that of another male but often over-runs territory occupied by several females. The latter bear up to three litters a year, each comprising up to five young, normally two or three.

The long-nosed bandicoots (genus *Perameles*) which, like their relatives, are nocturnal by habit and have a mixed flesh and vegetable diet, and the female's gestation period is about two weeks. The eyes of the newborn babies open after some forty-five days and they leave the mother's pouch at the age of nine or ten weeks, a comparatively short period for a marsupial, indicating rapid development.

The pig-footed bandicoot (*Chaeropus ecaudatus*) may now be extinct and little is known of its habits. It is basically vegetarian and probably breeds during the winter. The toes of this animal are reduced to an even greater extent than in related species, for only the second and third toes of the forefeet are well developed. The fourth is atrophied and the first and fifth are absent. The footprints thus resemble those of a pig, hence the common name. On the hind feet only the fourth toe is developed and serves as a point of support, the others being much reduced, the first non-existent. So the pig-footed bandicoot supports its body in a curious way. The forefeet are similar to those of artiodactyls, the hind feet to those of perissodactyls, the only difference being that the toes on which the body is supported are not identical. Unfortunately, this curious bandicoot is now extremely rare and in fact the last reported sighting was as long ago as 1926.

The bilby or rabbit bandicoot (genus *Macrotis*) has long ears, a slender build and a somewhat fragile appearance. The animals live in burrows with a single entrance, digging the holes with astonishing speed. The tunnels spiral down to a depth of 3–6 feet. Solitary by habit, each bandicoot appears to possess a piece of territory and it has been noted that within a radius of some 180 yards from each burrow there is no sign of another animal of the same species. The animals are carnivorous, feeding chiefly on insects and termites.

Most marsupials have little or no social organisation and this is particularly evident among bandicoots.

Among a multitude of small flesh-eating marsupials that look like rodents are the narrow-footed marsupial mice of the genus *Sminthopsis*, of which there are about ten species, some terrestrial, others arboreal. Food consists mainly of insects but they also hunt mice and lizards. The species *Sminthopsis crassicaudata* (*above*) breeds in June and July, a typical litter consisting of up to six babies.

### MARSUPIAL NATIVE-CATS

Class: Mammalia
Order: Marsupialia
Family: Dasyuridae

#### EASTERN NATIVE-CAT
(*Dasyurus viverrinus*)

Length of head and body: 14–18 inches (35–45 cm)
Length of tail: 8½–12 inches (21–30 cm)
Weight: male about 2½ lb (1.13 kg)
female about 1½ lb (0.68 kg)
Diet: mice, rabbits, birds, reptiles, amphibians, fishes, insects, carrion
Gestation: 8–14 days
Number of young: 4–8, exceptionally up to 24, but most of these die soon after birth.

Two colour phases, unrelated to sex; some greyish-brown with white spots, others dark brown with white spots, the latter the more common. The white spots generally sparse on head and limited to body. No first toe on hind foot.

#### WESTERN NATIVE-CAT
(*Dasyurus geoffroii*)

Length of head and body: 12–18 inches (30–45 cm)
Length of tail: 10½-14 inches (27-35 cm)
Weight: female about 1¼ lb (0.55 kg); weight of male not known
Diet: small mammals and birds, lizards, fishes, insects
Gestation: 8–14 days

Upper parts greyish-green with pinkish reflections, underparts white. Face lighter than other parts of body with numerous small white spots. Tail greyish-green, unmarked, the last third black. First toe of each foot lacks pad on the sole.

The smallest of all marsupials are the flat-headed marsupial mice of the genus *Planigale*. Although they weigh only about one-quarter of an ounce they are incredibly ferocious and will attack animals much larger than themselves, such as grasshoppers.

# The small carnivorous dasyures

The largest group of Australasian flesh-eating marsupials consists of animals belonging to the family Dasyuridae, ranging in size from that of a mouse to that of a dog. Most of these dasyures are small predators that feed on insects and other arthropods as well as on small vertebrates, especially rodents. Because of their appearance they are often called marsupial mice and marsupial rats, although they have a more pointed snout than the placental mice and rats of the family Muridae.

Unfortunately we know comparatively little about the biology and behaviour of these animals. Zoologists believe them to be extremely primitive mammals and certainly they are all agile hunters, with enormous appetites. Judging by observations of zoo animals, some of them are capable of consuming in one day a quantity of food exceeding their own body weight.

Many dasyurids exhibit highly interesting adaptations, as, for example, the flat-headed marsupial mice of the genus *Planigale*. These are the most diminutive members of the family, the length of head and body being 2–4 inches and their weight less than one-quarter of an ounce. The skull is very flat, exceptional for a mammal, with a minimum thickness of about 5 mm. This flattened type of head is more representative of snakes and may be an example of convergent evolution. Like snakes, these tiny marsupials often seek shelter in narrow rock fissures.

Despite their size the dasyurids are extremely ferocious and brave but it would be logical to suppose that their range of prey is very restricted. Yet these fierce little hunters are in fact capable of killing animals considerably larger than themselves, including grasshoppers.

Equally interesting are the desert-dwelling jerboa marsupials of the genus *Antechinomys* and they too illustrate the mistakes that a casual observer can make by trying to deduce the habits of an animal from its appearance. They closely resemble the jerboas of Africa and Asia, so much so that when first discovered they were assumed to be mammals belonging to the family Dipodidae, getting about in the same manner, namely by using their long hind legs to make huge jumps. But the fact that in these animals the marsupium of the female opens towards the rear clearly indicates that they do not move around in an upright position like the kangaroos, for if this were so the pouch

**LITTLE NORTHERN NATIVE-CAT**
(*Dasyurus hallucatus* and *D. albopunctatus*)

Length of head and body: 9½–12 inches (24–30 cm)
Length of tail: 8½–12½ inches (21–31 cm)
Diet: small vertebrates, insects, molluscs
Gestation: 8–14 days
Number of young: 6–8

Upper parts greyish-brown to dark reddish-brown; underparts brownish, yellowish or white. Upper parts of both species have white spots. Hairy tail, without spots and for three-quarters of length coloured like the back; lower side of tail and tip dark chestnut or black. First toe of hind foot present.

**TIGER CAT**
(*Dasyurus maculatus*)

Length of head and body: 16–30 inches (40–75 cm)
Length of tail: 14–22 inches (35–55 cm)
Weight: 4½–6½ lb (2–3 kg)
Diet: mammals, birds, insects
Gestation: about 3 weeks
Number of young: 4–6

Upper parts chestnut-red or dark chestnut; underparts pale yellow or sandy. White spots of variable size scattered over back, flanks and tail. Largest of marsupial cats, only one with spotted tail. First toe of each foot present.

*Facing page:* The narrow-footed marsupial mouse (*Sminthopsis murina*) is a ground animal but is capable of jumping high in the air to catch flying insects.

Eastern native-cat
(*Dasyurus viverrinus*)

Central Australian
jerboa marsupial
(*Antechinomys spenceri*)

Numbat or
Banded anteater
(*Myrmecobius fasciatus*)

Tiger cat
(*Dasyurus maculatus*)

Pig-footed bandicoot
(*Chaeropus ecaudatus*)

Eastern barred bandicoot
(*Perameles bougainville*)

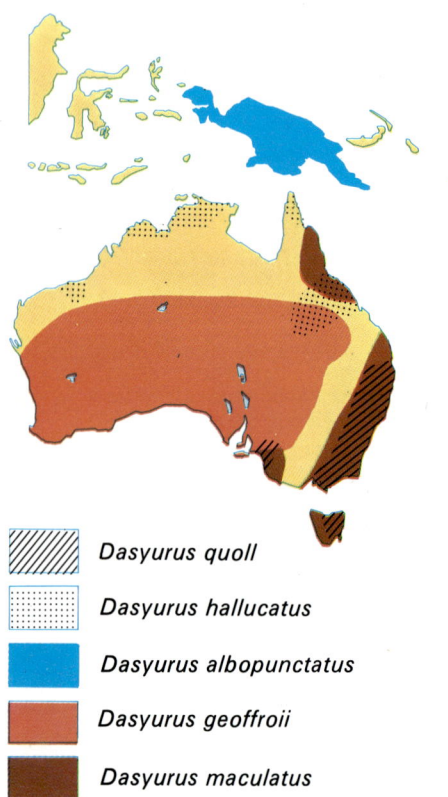

Dasyurus quoll
Dasyurus hallucatus
Dasyurus albopunctatus
Dasyurus geoffroii
Dasyurus maculatus

Geographical distribution of Australasian marsupial cats.

would be expected to open towards the front. High-speed photographs, together with detailed examination of footprints, show that these curious marsupials move in a kind of gallop, supporting the body alternately on the fore and hind feet and using the long tufted tail both for balancing and steering. The only times when they sit up on their hind legs, making use of the tail as an additional support for the body, is when they want to take a good look at their surroundings.

## Australia's native-cats

Other members of the Dasyuridae are considerably larger than the aforementioned species and are notable for their spotted coats. The first settlers of the continent gave them the name of cats but added the qualifying word 'native' to distinguish them from true cats. In fact all that these marsupials have in common with the Felidae is that they are carnivores and that they are approximately the same size. Actually they look more like members of the weasel family than cats.

These larger dasyurids are also agile and effective hunters. Species exhibit different types of dentition, reaching a peak of specialisation in the tiger cat or large spotted-tailed native-cat (*Dasyurus maculatus*) whose tooth structure is that of a typical carnivore. This is the largest of the native-cats weighing $4\frac{1}{2}$–$6\frac{1}{2}$ lb. As a tree-dwelling predator it occupies an ecological niche equivalent to that of the martens of other regions and possesses

the manoeuvrability and intelligence of placental carnivores. Despite their ferocity these animals are not difficult to tame, tolerant of human beings even highly companionable.

The tiger cat is an inhabitant of eastern and south-eastern Australia. Related species are the eastern native-cat (*Dasyurus viverrinus*), a resident of the east coast of Australia and Tasmania, and the western native-cat (*Dasyurus geoffroii*), with a wide range through central Australia and the west and south coasts.

## The vanishing thylacine

The largest of all flesh-eating marsupials is the thylacine or Tasmanian marsupial wolf (*Thylacinus cynocephalus*), better known in Australia as the marsupial tiger. Here is a truly astonishing example of convergent evolution for although these animals resemble wolves and other Canidae, closer examination reveals many features that separate them entirely from the true wolves which are of course placental mammals. Thus the ears of the former are more rounded and the muzzle much more deeply cleft (extending almost from ear to ear) enabling the animal to open its jaws in an enormous gape. Another distinctive feature is the pattern of 13–19 black stripes, hence the Australian preference for 'tiger', extending from the centre of the back over the rump to the root of the tail, possibly a means of camouflage. The rump and tail are also unlike those of true wolves but similar to those of kangaroos, the hindquarters tapering gradually

The broad-footed marsupial mice of the genus *Antechinus*, some species of which are arboreal, are agile nocturnal hunters, feeding principally on vertebrates and insects. They in turn are sometimes attacked and killed by nocturnal birds of prey, snakes and large carnivorous marsupials. Information about their behaviour is scanty but apparently they reproduce in August, giving birth to a litter of between three and eight tiny babies that are suckled for about three months and become sexually mature in approximately ten months.

and slightly curved, the long tail very powerful, flattened from side to side and so stiff that scientists who first studied the species claimed that if a thylacine were grasped firmly by the tail it would be incapable of turning around to bite. The hind legs are also distinctive for the tarsus is proportionally shorter than in the Canidae.

Once again it is sad to report that hardly anything is known about the behaviour of this species because the thylacine is today on the verge of extinction, possibly already extinct. Until quite recently Australians living in regions frequented by thylacines were concerned only to devise ever more efficient ways of exterminating them. Much of the information that scientists do possess is therefore largely based on hearsay and applies to a situation that may no longer exist.

It would appear that the thylacine, when it roamed freely through Australia and Tasmania – it became extinct on the mainland probably several thousand years ago – was a solitary predator and that not more than two animals would hunt together. Possibly younger animals would join their parents on such expeditions but there was never any evidence of the kind of social collaboration that characterises wolves and other members of the dog family. The thylacine was not as fast a runner as the true wolf and probably trotted long distances in pursuit of prey with a view to eventually exhausting them. Victims must have consisted of kangaroos, wallabies, smaller mammals and birds.

Ancient legends continually emphasise the fact that these carnivorous marsupials had a particular craving for blood and that they would try to sever the neck arteries of their victims, feeding first on the blood, leaving most of the rest of the body untouched. There seems to be no hard factual basis for such stories which probably do no more than add to the wealth of

The wolves and other Canidae, widely distributed through Eurasia, North America and Africa, are unrelated to the thylacine of Tasmania (now on the verge of extinction). Zoologists regard the superficial similarities as an interesting example of convergent evolution. Yet there are differences. Apart from the coat colour and pattern, the marsupial wolf has shorter, more rounded ears and a sloping rump more typical of that of a kangaroo. The shape of the head is also different and the jaws of the thylacine can open more widely.

*Facing page:* The marsupial cats of the Australasian region are medium-sized predators which occupy ecological niches similar to those of the small placental carnivores they closely resemble, the Mustelidae.

erroneous folklore, common to every part of the world, describing the allegedly evil habits of carnivores given to attacking domestic animals.

On the contrary, it seems to be generally agreed that the thylacine has never been guilty of attacking humans and apparently the only report of such an incident related to an abnormal animal which, half blind and almost starving, bit the hand of its mistress. In similar circumstances an ordinary dog might have done the same, so this proves nothing.

Another belief that has now been discredited is that the thylacine jumps like a kangaroo, but it is possible that it only rears up on its hind legs and tail to survey the surroundings, in the same manner as other marsupials.

The thylacine is undoubtedly a strong, courageous animal for reliable eye-witness reports have described the way in which it defends itself against packs of dogs. An individual dog, no matter how large, will find it more than a match. In one such encounter a particularly bold dog had a piece of its skull ripped off by a thylacine.

It is said that thylacines live in caves and rock clefts in inaccessible regions, sleeping during the day. Dr Guiler has suggested that this might apply only to breeding adults and their two to four young.

The conclusion of modern authors is that thylacines were once opportunistic hunters, attacking easy prey and feeding on carrion, thus playing a predatory role similar to that of hyenas on a continent where there are no large scavengers.

## Campaign of destruction

When Australia was first colonised the thylacine was widely distributed across the continent. The reasons for its subsequent extermination remain obscure but many authors believe that it was chased from its traditional haunts by the dingo. Discovery of fossil remains on New Guinea in 1960 show that at one time it inhabited that large island as well.

When Tasmania was settled by the white man thylacines were still present in large numbers, chiefly being found on the tree-covered savannahs. The colonists set fire to large areas, mainly with a view to clearing land for livestock and this inevitably proved a disaster for many indigenous animal species. Deprived of its habitual forms of prey the thylacine was compelled to switch its attacks to the only types of animal then available, namely lambs. Having discovered how easily such prey could be taken it stepped up its assaults on domestic animals, naturally arousing the wrath and enmity of farmers who immediately set about destroying it by every means at their disposal, a situation exploited by fur hunters who added to the general slaughter with poisoned bait. The entire campaign of extermination was given a fillip when the Van Diemen's Land Company instituted a reward system in 1840; and in 1888 the government itself offered one pound for every adult and ten shillings for every young animal killed. Official statistics show that between 1888 and 1909, the year in which this reward system was abolished, 2,268 thylacines

---

**THYLACINE**
(*Thylacinus cynocephalus*)

Class: Mammalia
Order: Marsupialia
Family: Thylacinidae
Length of head and body: 40–43½ inches (100–110 cm)
Length of tail: 20–26 inches (50–65 cm)
Diet: kangaroos, small mammals and birds
Number of young: 2–4

Outwardly resembles Candiae, but has more rounded ears and widely cleft mouth. Hind legs unusual with very short tarsi, tail very strong at base, flattened from side to side. Pelage light red; upper parts yellowish-grey or yellowish-brown with 13–19 brownish-black transverse bands.

**TASMANIAN DEVIL**
(*Sarcophilus harrisii*)

Class: Mammalia
Order: Marsupialia
Family: Dasyuridae
Length of head and body: 20–32 inches (50–80 cm)
Length of tail: 9–12 inches (23–30 cm)
Weight: male 12¾–20 lb (6.35–9.07 kg)
female 10–12¾ lb (4.53–5.44 kg)
Diet: carrion and wide range of live prey
Number of young: up to 4
Longevity: 7–8 years

Outwardly resembles a small hyena; short, massive head. No first toe on hind feet. Colour black or brownish-black with white band on throat and usually two white bands on flanks and legs. Pinkish-white muzzle.

were slaughtered, although the real figure may well have been considerably higher. Yet in spite of this bloodthirsty campaign the future of the species did not appear to be seriously threatened until 1910 when for some mysterious reason the population suddenly dwindled, possibly as a result of an epidemic, for at about the same time the numbers of Tasmanian devils also suffered a decline.

From that time onward the thylacine population was systematically driven from natural habitats and forced to take refuge in the eastern rain forests, surroundings which were alien and quite inappropriate given the fact that it was unable to find suitable prey in these areas. It became increasingly rare and was sighted only very occasionally.

Finally, but unhappily much too late, the Tasmanian authorities awoke to the fact that this animal was of scientific interest and decided in 1936 to take steps to protect the disappearing species. Since then there has been no reliable information as to how the dwindling population has fared, but it is obvious that the animals have been fighting a losing battle in a hostile environment. Certainly the prospects are bleak. It may not be long before the thylacine will be found only in the pages of literature, if indeed this is not already the case.

There have been a number of expeditions to discover whether the thylacine still survives. Not one of them has actually reported seeing a live animal although some claim to have found tracks. In 1961 a young male was accidentally killed on the west coast of Tasmania, encouraging the conservationists to continue

Tasmanian devils, despite their name, are not as ferocious and bloodthirsty as was once alleged. They are easily tamed and, provided conditions are favourable, are friendly and sociable.

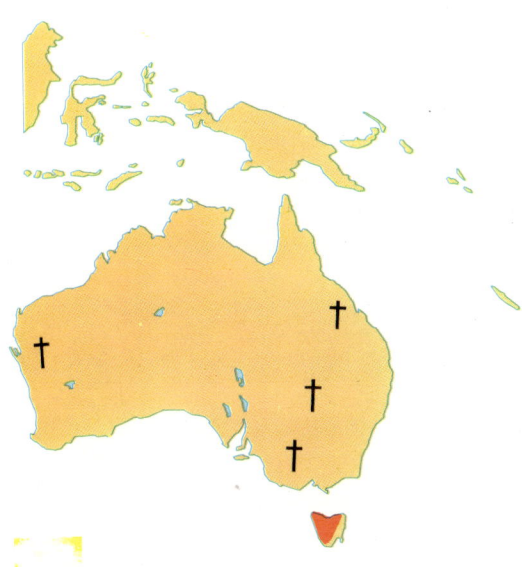

Geographical distribution (past and present) of the Tasmanian devil.

their propaganda work on behalf of the species. Their efforts were rewarded in 1966 when a large area of some 2,500 square miles in south-west Tasmania was set aside as a nature reserve. It is believed that there are still a few thylacines in this reserve as well as other animals such as the echidna, the platypus and the rare Tasmanian ground parrot.

Zoologists just do not know whether there is any future for the thylacine. Some optimistically look forward to a new growth of population that will enable scientists to begin their studies afresh; others are more pessimistic, believing that even if a few individuals have been spared they are too scattered and incapable of coping with their hostile environment. It is all too likely that the thylacine will remain as yet another testament to man's tragic irresponsibility.

## A 'devil' that belies its name

The thylacine is not the only large predator to have been driven from its original Australian habitat. A member of the Dasyuridae, the Tasmanian devil (*Sarcophilus harrisii*), would seem to occupy a position halfway between that of the thylacine and the native-cats.

The reason why this marsupial received its unflattering name is apparently due to the fact that the species has a reputation for ferocity, but this opinion was based on observation of the

The stocky, powerful Tasmanian devil is perhaps the most adaptable of all predatory marsupials, forced out of its original Australian habitat by man and his imported animals, and now restricted to Tasmania.

*Facing page*: Ecological distribution of some of the most typical Australasian marsupials. 1. Marsupial mouse. 2. Wombat. 3. Wallaby. 4. Tree kangaroo. 5. Jerboa marsupial. 6. Brush-tailed possum. 7. Koala. 8. Pig-footed bandicoot. 9. Cuscus. 10. Numbat. 11. Rock wallaby. 12. Sugar glider. 13. Marsupial mole. 14. Brush-tailed rat kangaroo. 15. Pygmy glider. 16. Greater glider. 17. Red kangaroo. 18. Grey kangaroo. 19. Tiger cat. 20. Tasmanian devil. 21. Wallaroo. 22. Thylacine. 23. Long-nosed bandicoot.

The broadly-based diet of the Tasmanian devil ranges from small wallabies, rat kangaroos and birds to reptiles, amphibians, fishes and crabs.

Geographical distribution of banded anteater.

---

**NUMBAT**
(*Myrmecobius fasciatus*)

Class: Mammalia
Order: Marsupialia
Family: Dasyuridae
Length of head and body: 7–11 inches (17.5–27.5 cm)
Length of tail: $5\frac{1}{4}$–$6\frac{3}{4}$ inches (13–17 cm)
Weight: $9\frac{3}{4}$–19 ounces (275–540 g)
Diet: mainly termites
Number of young: 4

Long, slender snout, small mouth, long, mobile tongue. Large eyes, small, pointed ears. Long tail. Five toes of forefeet and four toes of hind feet possess strong claws. Front part of body greyish-brown with some white hairs in subspecies *Myrmecobius faxiatus fasciatus*; subspecies *Myrmecobius fasciatus rufus* has brick-red pelage. Both have 6-7 white stripes on back, giving the effect of black and white cross striping. Black bands on cheeks encircling eyes with white streaks above and below.

---

behaviour of the first few individuals taken into captivity and evidently subjected to harsh treatment in deplorable conditions which would have infuriated the most peacefully inclined animals. Later studies showed that this initial judgment was misleading and that allegations of savagery were greatly exaggerated. Breeders were adamant in insisting that the animals they were used to handling were sociable by nature and never showed any aggressive traits even if somebody touched them while feeding, this being exceptional behaviour for a carnivore. Some even became devoted pets with exceptionally clean habits, bathing and washing the head with their forepaws.

With its black coat and white patches on throat, flanks and rump this strong, sturdy predator looks something like a small hyena. It displays a great measure of versatility for it can climb, swim and dive, this underwater prowess proving a valuable life-saving device. It is an opportunistic hunter and consequently highly adaptable. Prey may include wallabies, rat kangaroos, birds, lizards, amphibians, fishes and crabs; and sometimes it feeds on carrion. Apparently it has a keen sense of smell—a prime asset in its hunting operations.

Recently discovered fossils show that this marsupial was at one time an inhabitant of Australia but was also probably evicted by the dingo. Today it is found only in Tasmania where it is legally protected and apparently not greatly menaced for neither the fox nor the dingo have ever put in an appearance here.

The breeding season is in April and May and the babies, measuring about half an inch, are born towards the end of May or in June. At six weeks they have grown to 3 inches and continue their development inside the mother's closely sealed marsupium which contains four teats. They leave the pouch when they open their eyes and acquire a covering of hair, at about fifteen weeks. Soon afterwards the parents construct a rough shelter in a tree hollow or rock cleft. Alternatively, they take over the burrow of another animal. This provides refuge for the young which continue to suckle until five months old. The latter are sexually mature at about the age of two years.

## Marsupial anteaters and moles

If the thylacine provides an astonishing example of convergent evolution in which species of different orders and families from opposite sides of the world seem to develop along convergent lines in response to similar needs and challenges, the same phenomenon is apparent in the case of two other marsupials, the numbat and the marsupial mole.

The banded anteater or numbat (*Myrmecobius fasciatus*) is a rat-sized marsupial which specialises in hunting termites and its mouth parts are modified for this purpose. The snout is tapering, the tongue being long, mobile and sticky, darting in and out with amazing speed. There are a large number of teeth (about fifty) of varying shapes and sizes. All these features are present in other mammals of similar habit, such as South American anteaters, pangolins, aardvarks and echidnas except for the teeth. In most of these other 'ant-eaters' teeth are few in number and small or absent altogether.

Numbats live in tree cavities and unlike most marsupials they have diurnal habits. The species is in danger not only as a result of the predatory activities of dingos and foxes but also because of the gradual destruction of their traditional habitat.

These little marsupials are nimble climbers, exploring rotten wood infested by termites which, according to Calaby, constitute about 85 per cent of their food, the remaining 15 per cent being ants. To capture the tiny insects the numbat darts out its 4-inch tongue with lightning rapidity, if necessary ripping away the bark with its claws.

Four babies are born between January and March, apparently in a burrow dug by the mother. The absence in the female of a marsupium indicates the comparatively primitive stage of evolution of this species. The four teats are surrounded by a thick tuft of strong hairs to which the babies cling for protection and shelter.

Even more remarkable is the marsupial mole which surprised and perplexed the zoological world when first discovered in 1888. In fact the marsupial mole (*Notoryctes typhlops*) so closely resembles the placental moles that the American anatomist Cope was convinced of their relationship to the golden moles of the family Chrysochloridae. Only later was it established that

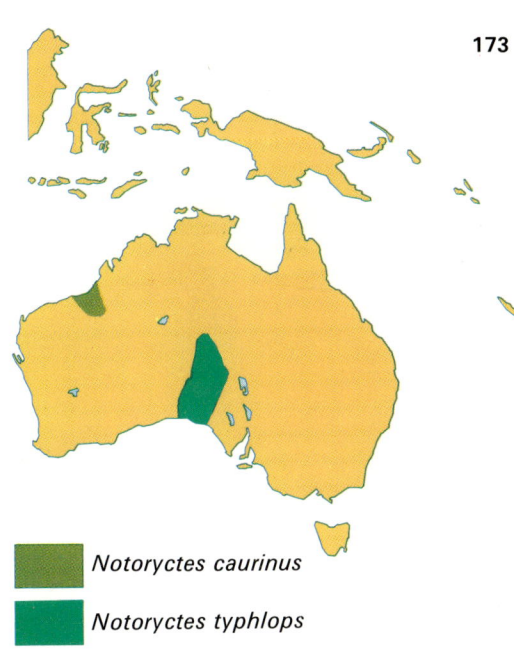

Geographical distribution of marsupial moles. (*Notoryctes caurinus* is now regarded as the same species as *Notoryctes typhlops*).

**MARSUPIAL MOLE**
(*Notoryctes typhlops*)

Class: Mammalia
Order: Marsupialia
Family: Notoryctidae
Length of head and body: $3\frac{1}{2}$–7 inches (9–18 cm)
Length of tail: $\frac{1}{2}$–1 inch (1.2–2.6 cm)
Weight: about $2\frac{1}{4}$ ounces (66 g)
Diet: insects, worms

Looks like placental mole with its rounded body, conical head, snout protected by horny shield, truncated tail. Eyes vestigial; ears mere openings, not visible through fur. Third and fourth toes of forefeet have long, spade-like claws. Soft, shiny, silky fur. Colour generally ranges from pure white to bright golden-red, with intermediate reddish tones.

The marsupial mole of the genus *Notoryctes* provides the most extraordinary example of convergent evolution. Lacking eyes or external ears this burrowing marsupial (1) bears an astonishing resemblance to the placental golden mole of the genus *Chrysochloris* (2), a resident of South Africa.

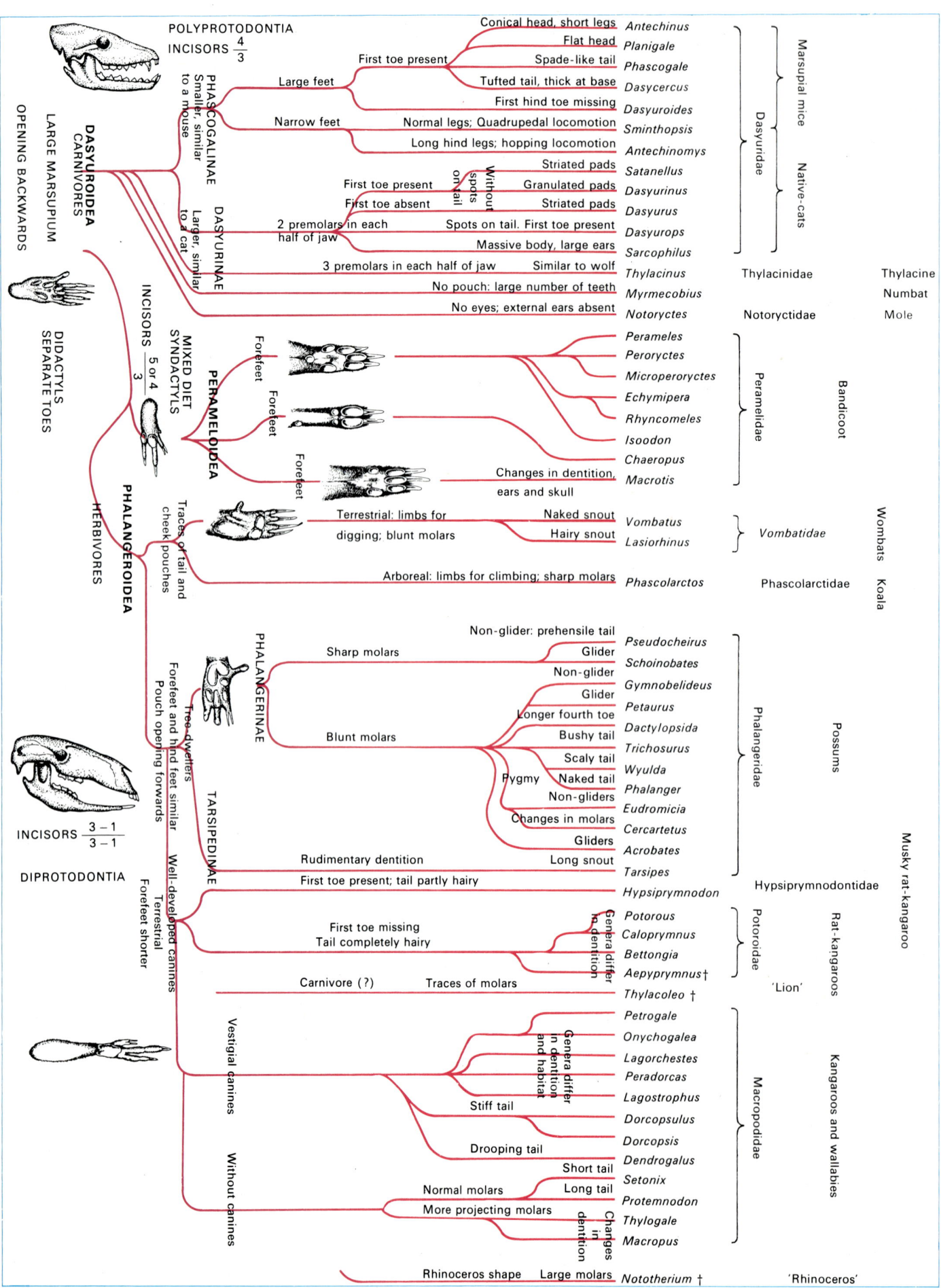

the animals were completely different and that their apparent similarities were yet another instance of convergent evolution.

The marsupial mole has a rounded body and a conical head with a snout protected by a horny shield. It has no ears, simply a pair of openings in the sides of the head. The eyes too, hidden by hair, are vestigial, barely 1 mm in diameter and without lens or pupil. The optic nerve is similarly reduced. The forefeet are furnished with strong claws, those of the third and fourth toes being enlarged and spade-like.

Hardly anything is known of these animals' habits. Observations of the few individuals kept in zoos show that they will pass from a state of complete lethargy into one of feverish activity, without any transitional phase. When awake they consume enormous amounts of worms and insects, lack of which may prove fatal. Burrowing like placental moles, they excavate to a depth of 3 inches or so but do not construct permanent tunnels, emerging at the surface from time to time, finding it easier to breathe above ground and indicating that they are not really acclimatised to a subterranean form of life. Often, before beginning to dig, they scamper about in a peculiar manner, leaving a triple pattern of prints corresponding to the marks made by their body and feet. When burrowing they work the tough little snout into the ground, shovelling away with the forepaws and tossing the excavated soil some distance behind them.

Information about breeding is scanty. Some authors believe that the female digs a burrow for the young when they leave the pouch.

The numbat or banded anteater has a mouth structure similar to that of other placental mammals specialising in hunting insects. This Australian marsupial feeds principally on termites and is feared to be in danger of extinction.

*Facing page:* This chart of the marsupials of the Australasian region, compiled by Sherwin Carlquist and based on the findings of a number of authors, outlines their evolutionary development, essential clues being teeth and limb structure. The evolutionary history of the animals of Australasia has been handicapped by the fact that there is very little documentation in the form of fossil remains.

# CHAPTER 102

# Watery oases of an arid continent

In every part of the world the freshwater wildlife, in lakes, marshes, rivers and streams, in their multiplicity of forms, provides naturalists with an inexhaustible field of study. This interest is heightened when they come to consider those rarer species of arid lands where the water so vital for survival is infrequently and sometimes only intermittently available in the form of lakes, swamps and pools which offer welcome relief in an otherwise monotonous landscape. The ornithologist may have to travel hundreds of miles across dusty plains before sighting a clump of dense vegetation which betrays the presence of water. But those dreary expanses of thorny scrub and those long hours of discomfort and thirst are quickly forgotten as he sets up his observation post among the tall plants fringing a marsh and settles patiently to watch the comings and goings of the birds whose activities have been briefly disturbed by his arrival.

The first bird he spots may well be the white-faced or grey river heron (*Ardea novaehollandiae*), somewhat larger than the little egret, and one of the first Australian species to be described by naturalists. This is a very trusting species, easy to observe as it fishes in shallow water or hunts grasshoppers in the tall grass, unperturbed by the presence of humans. Another bird common to these dry lands is the crested grebe, perhaps carrying its striped babies on its back. This delightful bird, well known in Europe, Africa and Asia, is a familiar inhabitant of the marshes of south-eastern and south-western Australia. Also present on all four continents, with a range extending to North and South America, is the great white heron; and here too, although much more sparsely represented, is the glossy ibis, unfortunately a fast disappearing species.

Although all these birds are fascinating to watch, the habits

*Facing page:* The elegant black swan, a familiar bird of Australia's lakes and ponds, is a close relative of the white swans that are similarly ornamental features of park and countryside in parts of the northern hemisphere.

Geographical distribution of the black swan.

The black-necked swan of South America has colour characteristics both of the black swan of Australasia and the white swans of northern climes.

of one species which is exclusive to Australia are of especial interest to the dedicated ornithologist. This is the black swan (*Cygnus atratus*).

## The black swan

The white plumage of the swans of Eurasia and North America has symbolic associations for many people, for these birds seem to epitomise purity and elegance of form. But Australia, that land of continual surprises, with mammals that lay eggs, boasts a swan that in appearance resembles its relatives of other regions in every detail, apart from the immediately striking fact that its plumage, instead of being white or whitish, is black. Only when the bird is in flight are the tips of the wings seen to be flecked with white. In a sense the black swan completes a colour cycle whose opposite extreme is the mute swan and whose intermediate example is the black-necked swan of South America.

Most of us have seen black swans in zoos or parks but how perplexed the first European colonists must have been to encounter a bird which looked familiar in all but this one important feature! Since that time the odd domesticated and imported white swan has escaped back to the wild, but until now only in New Zealand have wild colonies of the species really established themselves. Actually the black swan was introduced to New Zealand during the second half of the 19th century, a step which was soon regretted by the authorities for, as so often happens with imported species, the birds multiplied so rapidly that they caused extensive damage to cultivated land and incurred the wrath of local farmers. Nor were hunters any happier for the prosperity of the swan population seemed to coincide with a decrease in numbers of local game birds. In New Zealand today the species is on the decline as a result of hunting and nest destruction.

In its natural range of distribution the largest colonies of black swans are to be found in Tasmania and along the coasts of New South Wales, Victoria, South Australia and Western Australia. Nevertheless, the swans are present in smaller numbers

Black swan
(*Cygnus atratus*)

Black-necked swan
(*Cygnus melanocoryphus*)

Mute swan
(*Cygnus olor*)

in almost all wet regions provided that the water is less than about 3 feet deep, this being the maximum depth at which the birds can effectively immerse their long necks beneath the surface in order to feed on the bottom.

During prolonged periods of drought which often prevail in Australia many swamps dry up, compelling the swans to congregate in vast flocks (sometimes up to 50,000 strong) on those few that remain. The birds are extremely mobile and can fly long distances if conditions are favourable, although it is not accurate to speak of true migrations. Only prior to the moult, when their feathers are shed simultaneously and they are incapable of flying, will they head for the broad expanses of marshland where they can be certain of finding sufficient food to carry them through the difficult period when they are virtually defenceless and immobile.

There is no precisely defined breeding season and eggs are laid when local conditions are at their best. This flexibility is an asset for a water bird living on a continent where rainfall is spasmodic and irregular although, as is quite logical, most breeding pairs will site their nests in areas where the climate is reasonably stable.

Gregarious by nature, black swans construct their large nests (almost 3 feet in diameter) some distance from one another. The female lays five or six eggs which are incubated by both parents

The cygnet of the black swan acquires the red bill of the adult at about five months and the characteristic black plumage of its parents when it is a year old.

**BLACK SWAN**

(*Cygnus atratus*)

Class: Aves
Order: Anseriformes
Family: Anatidae
Wingspan: about 80 inches (200 cm)
Diet: aquatic vegetation
Number of eggs: normally 5–6
Incubation: about 40 days

Plumage black except for remiges which are white, visible only in flight. Red bill. Neck proportionately longer than in other swans. Cygnet born with greyish down; first feathers of young greyish-brown, becoming black after about a year.

Another difference between the black swan and the white swans of the north is that the former does not curve its long neck so markedly when adopting a threatening attitude.

for about forty days. The chicks are not all born at the same time and it is common to see one adult carrying babies on its back whilst the other partner is still incubating.

## The pink-eared duck

Another representative of the Anatidae which is exclusive to Australia is the pink-eared duck (*Malacorhynchos membranaceus*). In general appearance, although not in colour, it looks much like the shoveler of Eurasia and North America, for it too possesses an extra large, long, broad bill with numerous lamellae at the edges. This specialised type of beak is designed for obtaining food at the surface for it filters water so that only plants and microscopically small animals are retained. Because it feeds in this manner the duck adopts the position so characteristic of many other ducks, in which the front part of the body is submerged while the hindquarters remain above the surface a posture known as up-ending.

Since the pink-eared duck shows a preference for temporary pools and ponds formed after a fall of rain, it is a naturally nomadic species. Consequently it can benefit from a broadly-based diet. As soon as a storm creates puddles in the hollows of a plain these are immediately filled with all types of tiny creatures which quickly multiply, attracting flocks of these ducks which may become the most numerous species in areas where they may not have been sighted for years. When the pools run dry and the tiny animal inhabitants bury themselves in the mud in order to survive the ensuing period of drought, the ducks take wing once more and disperse to other areas.

Because the pink-eared duck is directly dependent on rainfall this is also a vital factor in its cycle of reproduction. Thus the birth of the ducklings coincides with the season when large quantities of rain create the ideal conditions for a colony to enjoy an ample supply of food. The adults often select a tree with a nest which has been abandoned by other birds, even if it is several feet above ground level; alternatively, they may simply nest in tree hollows. The eggs are generally laid from August to October in southern regions of Australia, March to May in northern areas. The female lays six to eight eggs which she proceeds to incubate for twenty-six days; and contrary to what happens among the majority of duck species, the male joins the female in defending the territory around the nest and also helps her to rear their progeny.

---

**WHITE-FACED HERON**
(*Ardea novaehollandiae*)

Class: Aves
Order: Ciconiiformes
Family: Ardeidae
Length: 24–30 inches (60–75 cm)
Diet: Insects, frogs, fishes
Number of eggs: 3–5
Incubation: 25–28 days

Colour blue-grey with white face and throat. Long tuft of feathers on head. Black remiges. Underparts greyish-brown. Bill about 4 inches long with white mark at base of lower mandible. Greenish or greyish band extends from base of beak to eye. Feet greenish-yellow which sometimes turn reddish in breeding season. Chick born with greyish down. Young has larger white facial mark and darker underparts.

---

*Facing page*: The pink-eared duck (*above*) and the Cape Barren goose (*below*) are exclusive to the Australasian region, the former widely distributed through Australia, the latter confined to the offshore islands in Bass Straits.

The pink-eared duck (*above*) and the wild duck (*below*), inhabitants of Australia and Europe respectively, have evolved the same type of bill, designed to filter water and retain only the plants and tiny animals on which they feed.

One of the important adaptations to life in the water which is exhibited by the platypus is the fold of skin which covers the eyes and ear openings when the animal is submerged. The toes of both forefeet and hind feet are webbed, the membranes of the former extending beyond the tips of the claws and folding back when the platypus is on land. The hind feet of the male are provided with venomous spurs, probably used for defensive purposes only.

*Facing page:* The platypus has caused many headaches for zoologists because of its many primitive anatomical features and its specialised adaptations to life in the water.

# Reptile, mammal or bird?

In the course of their long voyages of exploration across the uncharted waters of the Pacific the crew members of ships from Europe would frequently return home, after putting in at foreign ports, bringing strange gifts and purchases as souvenirs of their travels. Sometimes they would astonish their relatives and friends with stuffed or dried animals, notable either for their unusual shape or outstanding beauty. In the majority of cases such animals were examples of native species from the Orient or from Australasia and consequently completely unknown to European scientists. From time to time a sailor would bring back what he fondly hoped might prove to be a real prize—an animal which had been described to him as the legendary mermaid itself—with the body of a mammal and the tail of a fish. But when, in all innocence, the owner of such an animal presented it to a museum for closer inspection he was speedily disabused, realising too late that he had been duped by a cunning taxidermist who had tacked on the tail of a large fish to the skin of a monkey, so skilfully that all seams were invisible.

We have to remember that in the eighteenth century zoological knowledge was far less complete than it is today. At that time the known species of animals were numbered in thousands rather than hundreds of thousands, as they are today. So scientists had to guard their own inexperience as well. Thus they were doubly suspicious of anything unusual.

When in 1798 the skin of yet another mysterious animal was sent from Australia to London the zoologists were on their guard. For this creature was surely the most fantastic ever seen in the West, so improbable that the immediate conclusion was that this must be the most obvious and outrageous fraud yet perpetrated. Unbelievably, it had the beak of a duck, the body of an otter and the tail of a beaver; according to the consigner it had been caught in a river somewhere in eastern Australia. Con-

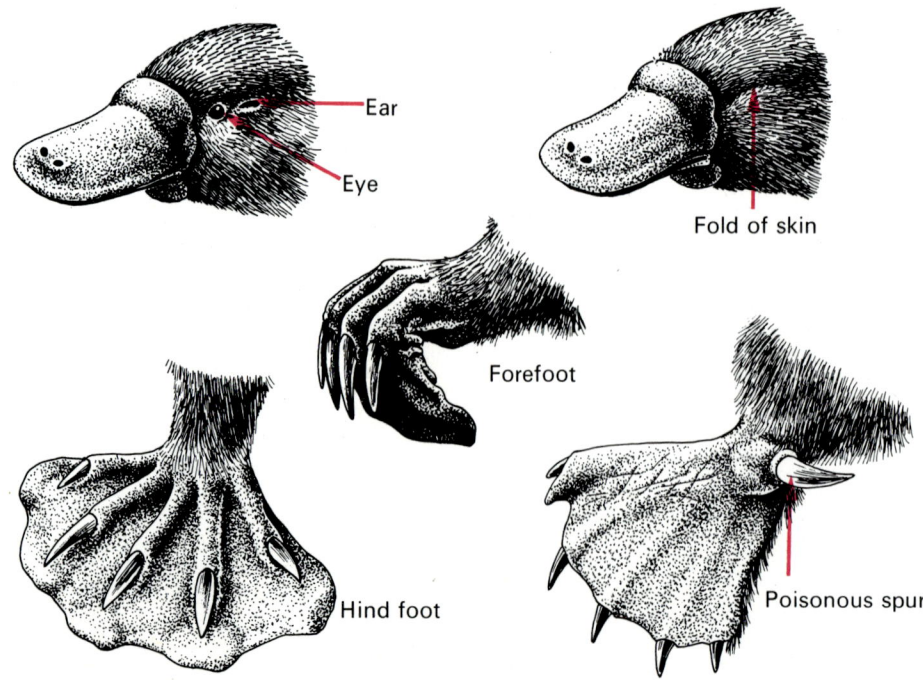

Geographical distribution of the platypus.

**PLATYPUS**
(*Ornithorhynchus anatinus*)

Class: Mammalia
Order: Monotremata
Family: Ornithorhynchidae
Length of head and body: 12–18 inches (30–45 cm)
Length of tail: 4–6 inches (10–15 cm)
Weight: 1–4½ lb (0.5–2 kg)
Diet: crustaceans, molluscs, frogs, worms, insect larvae, small fishes
Number of eggs: 1–3, usually 2
Incubation: about 10 days

Dark brown back, greyish or yellowish underparts. Thick, soft fur and sparser, longer hairs. Beaver-like tail; short, strong legs; large feet each with five webbed toes, and powerful claws. Long beak-like muzzle covered with naked skin and furnished with sensitive nerve endings. No external ears. Male has spurs on hind feet linked to venom glands, these spurs being present in young of both sexes but degenerating in female. Young possess calcified teeth but adults only horny plates. Young measure about 1 inch when born and are naked and blind. They leave the burrow at about four months, when they measure about 13 inches.

vinced that they were dealing with another common hoaxer, the naturalists at the receiving end used every means at their disposal to uncover the fraud (the marks of their labours are in fact visible at the base of the beak of the individual which was subsequently preserved in London's Natural History Museum). But on this occasion they were surprised and elated to find that they had something priceless on their hands, a zoological marvel constituting a find of major importance, detailed study of which was to herald a scientific revolution.

The outward appearance of this newly discovered animal (which scientists temporarily dubbed *paradoxus*) was astonishing enough; but detailed examination of its anatomy by Sir Everard Home in 1802 caused even greater surprise and confusion. Thus study of the reproductive, urinary and genital apparatus revealed that instead of possessing separate external outlets, as in mammals, these tracts were all linked to a common cavity known as the cloaca, characteristic of birds, reptiles and amphibians. Furthermore, certain portions of the skeleton, as, for example, the joints linking the forelegs with the trunk, resembled the joint structure of lizards in certain details. The internal temperature of the animal, fluctuating between 25°C (77°F) and 30°C (86°F), also indicated an imperfect system of temperature control, placing it somewhere between the homoeotherms (animals with constant body temperature) and the poikilotherms (animals with variable body temperature), that is, between birds and mammals on the one hand and reptiles on the other. The discovery that the female of the species possessed mammary glands seemed to indicate clearly that some kind of mammal was involved, but certainty turned to doubt and then to utter perplexity when the Australian zoologist W. H. Caldwell extracted a soft-shelled, whitish egg from the cloaca of an adult female.

After several years of discussion and dispute zoologists came to the conclusion that this amazing animal with a duck's bill, otter's body and beaver's tail, laying eggs like a reptile and yet suckling its offspring, was indeed a mammal; but undoubtedly it was unlike any mammal yet known to science. There was no alternative but to place it in an order of its own to which naturalists gave the name Monotremata ('single orifice'). It came to be known as the platypus or duckbill (*Ornithorhynchus anatinus*).

Because of its many primitive characteristics—the cloaca, the egg-laying method of reproduction, the imperfect temperature-regulating system and the absence of teats—the duckbill or platypus is virtually a living fossil, representative of a branch of animals which disappeared from the earth millions of years ago, being superseded by the more highly evolved placental mammals. Its survival on the Australian continent alone is due to the fact that it has been spared competition for food and living space by reason of geographical isolation over the centuries. Nevertheless, the platypus is no mere intermediate rung in the evolutionary ladder which has reptiles at the bottom and mammals at the top, but a completely distinct and independent species which has diverged from the main line of mammalian evolution.

The beak of the platypus has often been compared with that

of a duck because of its general shape; but in fact a comparison with the muzzle of a dog would be more apt, for it is a fleshy structure, covered with naked skin, always damp, and furnished with a large number of nerve endings which make it a highly sensitive organ.

Both the hands and feet of the platypus are webbed, helping the animal to move effortlessly through water and to walk on muddy or swampy terrain. The membrane of the forefeet, which are used to propel the animal when swimming, is very large, extending beyond the tips of the toes. This also provides a broad supporting surface on land. As soon as the platypus emerges from the water the web folds back, leaving the claws free for walking or for digging. The hind legs of the males are equipped with a pair of horny spurs linked to venom glands. The poison ejected by the platypus is very weak, capable of causing a man severe pain, but not fatal. The spurs puncture the skin while the platypus lashes out backwards with its feet. This poison mechanism would appear to be designed solely for defensive purposes and not for hunting. Although superficially the possession of spurs and venom glands might seem to imply a relationship with reptiles, this is not the case for the structural details are entirely different.

It might be permissible, however, to recall without making too strong a point of it, that some of the reptiles of a hundred million years ago, in the Age of Reptiles, had spurs, usually on the forelegs, the function of which is not fully known.

Although the platypus is well adapted for aquatic life it spends much of the day on land and only a few hours each day

With its eyes and ear openings concealed by a fold of skin, the platypus probes with its sensitive bill for underwater prey.

Typical animals of the watery habitats of the Australasian region. 1. Johnson's crocodile. 2. Mudskipper. 3. Marsh harrier. 4. Whistling tree duck. 5. Chestnut teal. 6. Black swan. 7. White-faced heron. 8. Little egret. 9. Glossy ibis. 10. Great crested grebe. 11. Platypus. 12. Saltwater crocodile. 13. Water monitor. 14. Long-necked turtle. 15. Pink-eared duck. 16. Pygmy goose.

in the water. At dawn and at dusk (and at other times of day provided the sky is overcast) the platypus leaves the burrow it has excavated in the stream or river bank and enters the water, swimming on the surface, using the forefeet as propellors and the hind feet and tail as rudders and stabilisers. Only the muzzle, the top of the head and the back protrude above the water. Now and then it dives right under and when it does so the eyes and ear apertures (there are no proper ears) are sealed by folds of skin. Blind and deaf below the surface, the platypus probes among the stones, pebbles and sand of the river bed with its sensitive snout, searching for small crustaceans, worms, larvae of aquatic insects, freshwater snails and other tiny water animals. Prey is then either swallowed or stored in the cheek pouches.

A platypus can stay submerged for five minutes or so but does not usually remain below for more than a minute or two. It then floats to the surface in order to breathe and consume whatever it has caught.

Young platypuses possess genuine calcified teeth but the adults do not retain these, their teeth being more like horny plates. The absence of proper teeth makes chewing difficult and this is perhaps why the animal scoops up a little sand from the bottom whilst feeding, as an aid to mastication.

The possession of horny plates in the jaws of an adult mammal is a most remarkable feature. It is, however, not a primitive feature but a highly specialised one. It may be that the platypus' diet once consisted mainly of a food, now extinct, for which horny plates were eminently suitable.

When its large appetite is satisfied, the platypus returns to

its riverside burrow—a tunnel between 12 and 30 feet in length, but sometimes much longer, up to 100 feet—the entrance or entrances being situated some 3–6 feet above the water surface, although low entrances may be covered during floods. The tunnel is not dug in a straight line but contains many turnings. This is home for a pair of platypuses who live together throughout the year except during the breeding season, when the female digs her own. Breeding occurs during the Australian spring but commences at different times, depending on latitude. In northern Australia the animals will mate in July or August whereas in the south sexual activity may continue until October.

Prior to mating (which takes place in the water) there is a courtship ritual. The two animals swim in circles on the surface, the male seizing the female's tail with his beak.

Shortly after the rut the female leaves the burrow and digs a new one, about 20–60 feet in length. At the end of it she prepares an oval-shaped chamber lined with leaves, and at intervals along the whole length of the gallery she places plugs of earth, heaped up with her flat tail. These barriers are opened and closed every time she goes in and out. It has been claimed that these mounds are a form of defence against potential predators of eggs and young; snakes and water rats, both common along rivers, are potential predators of the young platypus. Some authors have suggested, however, that it is a form of behaviour inherited from primitive ancestors of a bygone era when the platypus may have been menaced by carnivores that have vanished. This is an attractive theory but virtually impossible to prove. Possibly these heaps of soil also help to keep atmospheric conditions fairly stable inside the nest.

About two weeks after copulation the female lays from one to three whitish eggs in her leafy nest and incubates them under her body for approximately ten days. The babies are born naked and blind; and since the mother does not possess teats they lap the milk which oozes from the mammary glands on her abdomen. The young platypus does not leave the burrow and enter the water until it is seventeen weeks old.

Although these details are given with an air of confidence the fact remains that the observations from which they are derived are largely fragmentary and scattered. David Fleay alone has bred platypuses in captivity.

After mating, the female platypus leaves the burrow where she has lived with her partner for the rest of the year and digs a new one, to a length of 20–60 feet, at the end of which is the nest where she lays her eggs.

Towards evening, or in full daylight if the sky is cloudy, the platypus leaves its riverside burrow and ventures into the water to look for food.

Australian water rat
(*Hydromys chrysogaster*)

The function of the spurs on the male platypus is little understood. It has been suggested that the spurs are used in fighting, as weapons against predators, or for immobilising the larger prey. The poison apparatus consists of a movable horny spur on the inner side of each hind leg, near the heel. When attacking the human hand or wrist the platypus brings its hind legs together with considerable force so that the spurs are embedded in the flesh. The venom flows from a gland in the thigh. The injured hand begins to swell, and the swelling, which may later extend as far as the shoulder, lasts from one to a few days. Intense pain is experienced during the first day and this is followed by a soreness that lasts several days, or even for some weeks.

At the end of four months the young accompany their mother into the water for the first time and begin to feed on their own. At about two and a half years of age they are sexually mature. Platypuses in captivity live for ten years or more.

## The adaptable water rats

Although Australia is memorable for its strange and unusual forms of wildlife—the platypus, echidna, kangaroo and koala spring immediately to mind—it must not be supposed that the local mammals consist exclusively of marsupials and monotremes. In fact the great island continent is the home of many placental species, including rats and mice, bats and seals. More than half the number of animal families and almost the same proportion of genera represent placental mammals.

The presence of seals is hardly surprising because of their swimming abilities. As for the bats, they have had no difficulty in colonising Australia since they are flying mammals. Of the twenty-one bat genera in the Australasian region only three are truly endemic. The case of the rodents is different since it has been no simple matter for them to overcome the natural obstacles presented by bodies of water. Their invasion of the continent and the adjacent islands has been the consequence of a series of leaps forward from island to island (often as passengers aboard improvised rafts or clumps of floating vegetation) where they have gradually become differentiated. Of the thirteen genera of the region, ten are endemic.

All the Australasian rodents belong to the family Muridae, which after their arrival diversified so as to adapt to every available ecological niche. The most interesting individuals are unquestionably the aquatic rodents for they appear to have adapted to life in the water only after reaching Australian shores. The head of a typical water rat (*Hydromys chrysogaster*) is long and slender, the nostrils placed high on the snout so that the animal can breathe without raising its head above the surface, the eyes are positioned high in the head, the ears are small, the fur is short and compact (like that of seals) and the feet are broad and partially webbed.

Nocturnal by habit, the Australian water rats leave their burrows as evening approaches and venture out on their nightly quest for food. This consists in the main of crustaceans, snails,

frogs, fishes, water birds and their eggs. As a general rule, once they have captured their prey, they carry it to a rock or tree trunk jutting out of the water and eat it at leisure. One aspect of behaviour which long baffled zoologists was how the rodents managed to cope with bivalved molluscs. But one day a naturalist watched a water rat carrying a mollusc to a rock and simply letting it lie there in the sun until the two halves of shell were forced open by the heat. Other naturalists have seen rodents killing larger animals with a swift bite in the neck.

The habit of removing shellfish from the rivers and leaving them in the sun deserves special attention. It seems to be habitual, according to recent observations by G. J. Barrow of the Queensland Institute of Medical Research.

During the day water rats remain hidden in a hollow tree trunk on a pile of vegetation or even in the nest of a swan, building a small cavity to let the water flow out. But those living in rivers prefer to dig a burrow in the bank for their daytime shelter. At the end of the tunnel is a huge chamber which they line with leaves, bark and twigs. In the same burrow is another chamber filled with bones and shells, apparently indicating that

Australian water rats are useful allies to man because they feed on the aquatic snails that harbour parasites which infect sheep. Since they are hunted for their fur—a threat to their numbers—the authorities have introduced a close season for such activities during part of the year.

*Following pages*: After years of persistent rumours of monster lizards inhabiting lonely islands in the Far East, zoologists finally confirmed the existence of the Komodo dragon, largest of living monitors. Today these reptiles are protected on Komodo itself and two other small Indonesian islands.

it is used as a kind of storehouse.

When the breeding season approaches the water rats build another type of nest consisting of a large chamber from which leads a second tunnel that terminates in an opening above ground, concealed in the grass; the latter serves both as an emergency exit and a source of ventilation. The female gives birth to one or two litters annually, each comprised of a relatively small number of babies, four or five. Because of this low birth rate the size of the population has tended to decrease steadily, the situation being aggravated by man who hunts the animals for their precious fur. The Australian authorities have tried to institute a close season for hunting in order to control the hunting pressure on the species, for unlike many other rodents which are regarded as a scourge, these rats, which are adapted so wonderfully to life in water, are useful animals, consuming aquatic snails which act as hosts to parasites of sheep.

## Turtles of river and swamp

The laying of an underground telephone cable between two towns is a common enough operation which does not normally arouse public interest or concern. But when in 1967 it was planned to lay such a cable between Perth and Carnarvon in Western Australia there was an immediate protest from conservationists who pointed out that if the project went through it would gravely threaten the future of an animal species. As originally planned, a portion of cable was to be laid across a small piece of marshland—all that now remained of a once-enormous stretch of swamp, which had been drained and given over to farming. The crux of the matter was, however, that this small area was the home of a rare species, the short-necked swamp turtle (*Pseudemydura umbrina*), already threatened with extinction. To save it the conservationists had previously set aside a small zone of about 750 acres where some 200–300 of these turtles, legally protected, lived undisturbed. It was feared that any work involved in laying the telephone cable would inevitably harm the already sparse turtle community. Happily the appeals of the lovers of wildlife prevailed. The engineers' plans were modified (at considerable additional cost), the cable taken around by an alternative route, and a precious species saved.

The swamp turtle has the characteristic, common to all Australian turtles, of being incapable of withdrawing its head into its shell by folding the neck into a vertical S-shape, in the event of danger. The structure of the cervical vertebrae allows it only to bend the head sideways, wedging it as far as possible between the upper shell section or carapace and the abdominal shell or plastron.

There are some dozen species of Australian turtles, all of which belong to the family Chelidae. They are, however, subdivided into two distinct, easily recognisable groups, the short-necked and long-necked turtles. In the latter group the neck may be even longer than the rest of the body.

These turtles live almost exclusively in rivers and swamps and are often seen swimming on the surface or sunning them-

The side-necked swamp turtle has not disappeared thanks to the fact that its habitat, a small area of marsh in Western Australia, is now a protected nature reserve.

*Facing page:* The turtles of the Australasian region, like their relatives in the northern hemisphere, can retract their neck if danger threatens but the structure of the cervical vertebrae permits them only to bend it sideways, as can be seen here in the picture of the long-necked turtle (*above*). The short-necked swamp turtle (*below*) has been the object of detailed scientific study. Zoologists have fitted the carapace of selected animals with miniature radio transmitters so as to follow their movements when returned to the wild.

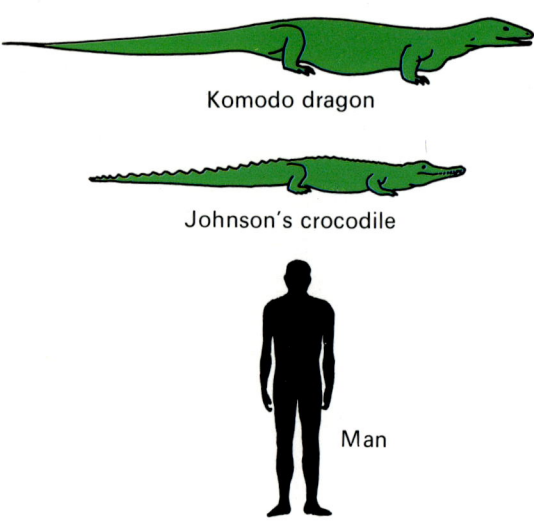

The huge size of the Komodo dragon can be appreciated when compared with that of Johnson's crocodile and man.

selves on a rock or a tree trunk sticking out of the water. But sometimes, when drought dries up their watery habitat, they abandon the marshes *en masse* and embark on hazardous cross-country journeys. Logical as it may seem for them to venture abroad in search of more favourable regions in times of drought, it is less easy to explain why groups of these turtles may leave the certain refuge and comfort of a river or lake and strike out across dry land on a journey which must end in death for many of them. It is not uncommon for these turtles to come up against a barrier erected by farmers to keep out rabbits. Faced with such an obstacle, the slow-moving turtles wander to and fro indecisively until they die from exposure to the searing rays of the sun, or from starvation.

## Crocodiles of tropical Australia

Palaeontologists working in the desert regions of the Australian interior have discovered the fossil remains of crocodiles, indicating that in times long past conditions in these arid regions were very different from those now prevailing, similar, in fact, to those existing today along the northern coastal belt. It is here, in northern Australia, that the only surviving crocodiles of the Australasian region live, in the tropical climate which suits them best.

There are two Australian species, the saltwater crocodile (*Crocodylus porosus*), already described, with a range of distribution extending to India, and Johnson's crocodile (*Crocodylus johnsoni*). The former is an inhabitant of coastal regions, estuaries and the lower reaches of streams and rivers, although when the water is exceptionally low it may make its way into the interior. Only very occasionally is this dangerous species encountered in lakes where the odd individual has perhaps been stranded by flood waters.

Johnson's crocodile is an inhabitant of rivers and freshwater lakes, feeding essentially on fish. Some individuals may measure more than 10 feet but most of them are smaller. Timid and retiring by nature, this crocodile poses no danger for humans, unlike the saltwater or estuarine species which has occasionally been known to kill people. Although inoffensive, Johnson's crocodile will defend itself vigorously when at bay.

The breeding season for Johnson's crocodile is at the end of the dry season when the female lays about two dozen eggs in a nest, approximately one foot deep, which she digs in a sand bank. The young crocodiles are born about eight weeks later, hatching shortly before the arrival of the monsoon rains which cause rivers to overflow and flood the banks where the eggs have been incubated.

## A forgotten giant

Lizards are found in all parts of Australia but the largest members of the group, the goannas, or monitors, are especially adapted to life in dry places. Two of the native species found near water are the mangrove monitor (*Varanus semiremex*), an inhabi-

*Facing page:* Johnson's crocodile is an inhabitant of rivers and freshwater lakes of northern Australia. It is a harmless species which feeds mainly on fish.

The large lizards known as monitors are found in many parts of Australia. The two typical aquatic species are the mangrove monitor and the water monitor, illustrated here, both of which are inhabitants of northern coastal regions.

tant of northern coastal swamps, and the water monitor (*Varanus mertensi*), which is also a resident of northern Australia, but chiefly found along lakes and rivers.

This family once included another large monitor, more than 15 feet in length, which evidently lived in Australia in the Miocene age but which disappeared from the earth long before man made his appearance.

Nevertheless, the giant of the family, not found in Australia itself but indisputably an animal of the Australasian region, survives to this day. It is known as the Komodo dragon (*Varanus komodoensis*), one of several large animals to have been discovered in the 20th century, decades after several eminent zoologists had prophesied that there were no more large animals remaining undiscovered.

The habitat of this enormous lizard is nowadays restricted to several small islands situated in the Sunda archipelago, just south of the Wallace Line that separates the Oriental and Australasian regions. Yet despite its enormous dimensions, this formidable representative of the monitor family, the largest ever known, was only discovered by herpetologists at the beginning of the 20th century. Prior to that, however, many rumours had circulated about these monstrous reptiles.

For centuries the inhospitable islands where these giant lizards lived remained uninhabited by man. From time to time pearl fishers and turtle hunters put in to anchor off their coasts and brought back tales of gigantic lizards they had seen basking on the beaches but such stories were dismissed as figments of the imagination. In due course, however, the sultan of the island of Sumbawa decided that the islands could be put to some use as prison settlements for criminals and other undesirable persons. The few men who survived and were set free now backed up the earlier tales, describing with relish the ferocious monsters they had seen, some of them measuring more than 25 feet.

Zoologists began to think that perhaps there was some basis of fact in these much-repeated tales. Eventually, in 1912, P. A. Owens, the director of the Buitenzorg Botanical Gardens in Java, requested the governor of Flores to send an expert to investigate the rumours. In the same year an officer of the Dutch colonial army visited the island of Komodo, hunting and killing a lizard which was about 10 feet long. Owens was satisfied that the earlier reports had been much exaggerated but was able to announce to the scientific world the hardly less exciting reality of the Komodo dragon.

The 1914–18 War turned men's minds to more important matters but some years after it ended zoologists returned to Komodo. In 1923 Duke Adolf Friedrich von Mecklenburg brought back to Europe the skins of four of the monitors; and three years later the Komodo dragon was studied in its wild habitat by the naturalist E. R. Dunn.

The population on Komodo was clearly on the decline. In 1930 there were still reports of individuals measuring 13–14 feet, but by 1963, with only 300 or so of the reptiles known to be surviving, most of them on Komodo, the largest individuals measured only about 10 feet. Although officially protected, the numbers were still decreasing, mainly as a result of local hunting. Evidently the flesh made good eating.

The Komodo dragon is a carnivore and examination of stomach contents has revealed heads of wild pigs, hooves of deer, horns of buffaloes, and remains of monkeys, in addition to rats, snakes, fishes and smaller lizards. The reptile usually catches prey by gliding noiselessly through the tall grass until it comes within range of a herd of grazing deer or wild pigs, or perhaps a troop of macaques. Then it pounces on its selected victim which is powerless in the grip of the monitor's enormous jaws. Apart from live prey the reptiles devour turtles' eggs on the beaches and may climb trees to filch the contents of birds' nests. They also eat carrion, locating the carcases mainly by smell (the forked tongue transferring odour particles to Jacobson's organ).

The Komodo dragon is a threatened species. The southern part of the island of Rincha and the uninhabited island of Padar have been converted into reserves; but on Komodo itself protection is difficult to enforce for the local tribes habitually set fire to the vegetation, destroying both hunter and prey.

---

**KOMODO DRAGON**
(*Varanus komodoensis*)

Class: Reptilia
Order: Squamata
Family: Varanidae
Length: up to about 10 feet (3 m)
Weight: 265–330 lb (120–150 kg)
Diet: insects, rodents, deer, pigs, monkeys, eggs, carrion, etc

Largest living monitor. Greyish-brown to reddish brown with darker circular marks. Body slightly flattened; tail as long as head and body. Short, powerful legs; long claws on toes. Long, slender, forked tongue. Young are darker than adults with reddish marks all over body and vertical greenish stripes on neck which later disappear.

# CHAPTER 103

# The oceanic islands: a panorama of evolution

From what we know today isolation is an indispensable prerequisite for the formation of species. If all the world's regions were inter-connected, with no natural barriers to hinder cross-breeding and the free circulation of genetic material, we would have one immense, homogeneous living species – a dismal prospect. Fortunately there have always been islands (in the evolutionary sense), created by mountain chains, rivers, forests, climatic factors and the like. The existence of such natural obstacles has meant that groups of animals have remained separate from one another, conserving their own forms and habits, eventually severing links, from the reproductive point of view, with older relatives from whom they are now geographically cut off by a wood, a mountain or a stretch of water.

Of course, the ideal conditions are created wherever islands, in the true geographical sense, exist. In lands encircled by thousands of miles of ocean, far from the highways of migration and commerce, access is only possible to those animal species that can fly (such as birds and bats) and to those which are hardy enough to endure long journeys on improvised rafts (such as small reptiles and the prolific, cosmopolitan rodents). Here they are likely to find a paradise of plant life where they can evolve in spite of the pressure of predators and not be continually subjected to competition for available habitats. Such species have flourished, giving rise to a multitude of varied and sometimes curious forms.

Colonisation of an island surrounded by an immense expanse of ocean is an adventure into the unknown; but in the case of islands situated nearer to continental land masses there is more predictability for isolation has occurred comparatively recently. The British Isles are a case in point. Such islands rest on con-

*Facing page:* Many of the world's islands are sanctuaries of wildlife where rare and primitive species have been preserved, thanks to having been geographically isolated for centuries and thus spared the competition of other animals. They have not, however, escaped the direct and indirect pressures brought about by human colonisation and cultivation and the future of many an island paradise is now imperilled. New Zealand, whose South Island is pictured here, is renowned for its natural beauty and wonderfully varied flora and fauna.

Three important waves of migration have contributed to New Zealand's mosaic of animal life. The first arrivals, shown here in green, date from the Mesozoic era, crossing by land links that have long since vanished. They were the ancestors of a primitive native fauna, some species of which still survive. Much later, when Polynesian man arrived (probably about a thousand years ago) he brought with him the rat and the dog, shown in black. The species shown in blue are some of those imported in large numbers after 1779, both to New Zealand and neighbouring islands, with serious and often disastrous consequences for endemic species.

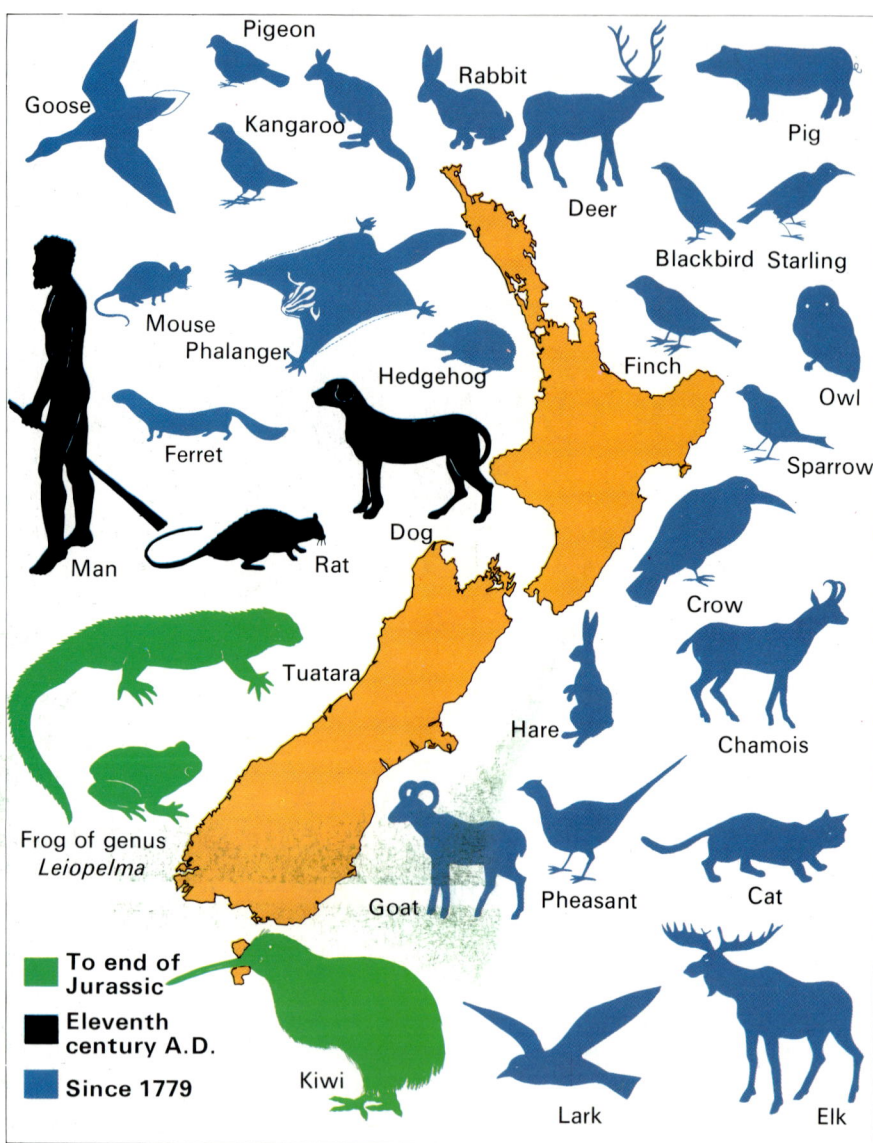

*Facing page:* New Zealand is a land of great natural beauty and striking scenic contrasts, including forest-fringed fjords, snow-capped mountains and volcanos, some of them still active.

tinental shelves and the plants and animals to be found there are not significantly different from the flora and fauna of the continents to which they were formerly linked. Certainly climatic and geological influences such as glaciations and volcanic eruptions may bring about important changes in the wildlife pattern of continental islands; but scientists now have the knowledge and equipment necessary to discover the relationships between the species concerned.

The situation is very different in the case of the oceanic islands which have been formed as a result of a submarine volcanic eruption or the slow, relentless activities of marine invertebrates such as coral. However formed, plant life has sprung up in the most unexpected ways. Thus vegetable seeds may reach these shores on rafts formed by their own impermeable shells, as, for example, the coconut; and both seeds and invertebrates may be transported in the excrement and feathers of migrating birds.

The smaller seeds do not necessarily require transport by water or by birds. Airborne seeds have been found well out to sea, a long way from the nearest land.

The vegetational cover of New Zealand has been devastated by indiscriminate tree-felling and deliberately caused fires. There remain, however, some untouched forest areas which include coniferous trees of the genus *Podocarpus* together with densely growing ferns and other low plants, relics of a prehistoric era.

The various marine birds which were capable of overflying immense stretches of ocean found many such recently emerged oceanic islands ideal for nesting purposes. But there were also certain common ground species such as finches or clumsy fliers such as rails which reached these shores, perhaps having been driven off-course from their customary migration routes by storms and hurricanes. For these birds an oceanic island might turn out to be a paradise or a graveyard, a fortuitous outcome depending on a number of factors—the size of the island, the climate, the availability of water, the extent of plant cover, and the presence of invertebrates on which to feed.

The animal inhabitants of islands which have been isolated for a considerable length of time and have thus been unaffected by outside influences and, in a manner of speaking, not subjected to the basic laws of evolution that shape continental species, possess a number of distinguishing characteristics. Thus when a group of animals first set foot on an island rich in vegetation, no matter how they have managed to do so, they find a variety of ecological niches available and proceed by stages to occupy them. This is what happened in the case of the birds discovered on the Galapagos Islands, subsequently named Darwin's finches, and the Hawaiian honey-creepers. The modifications of the bill exhibited in these species are excellent illustrations of the ways in which such birds, stemming from primitive ancestral forms, gradually acquired the anatomical features necessary for adapting to a specialised diet, whether of seeds, fruit, grass, pollen or insects.

When such island species have evolved over an exceptionally long period amid surroundings where vegetation has been abundant and ground predators non-existent, the birds have been free to settle in habitats which in other parts of the world would have been occupied as a matter of course by herbivorous mammals. The moas of New Zealand, the elephant birds of Madagascar (Malagasy) and the dodos of Mauritius, for example, originally had all the food they required in regions where no large herbivores roamed, where no competing species prevented their expansion and where no carnivores threatened their population growth. Such birds not only lost their capacity to fly (which they once must have been able to do in order to reach their destinations) but also attained massive dimensions, more characteristic of mammals.

It is for these reasons that oceanic islands are veritable sanctuaries of wildlife, harbouring rare, archaic species which, had they lived on a continent, would have been supplanted by later, more highly evolved species and doomed to extinction. They are also museums of native species, that is, of animals which, being descended from the earliest arrivals (themselves of continental origin), have subsequently developed specific features that are not found anywhere else in the world.

Yet some islands have proved to be cemeteries for certain endemic birds which have been unable to contend with newer species introduced by man or to withstand the depredations of man himself who, in the space of a few centuries, sometimes decades, has destroyed the peaceful isolation that has reigned for thousands of years. Such was the tragic fate of the moas, exterminated by generations of Polynesian hunters, and of the dodos, clubbed to death by sailors who filled the holds of their ships with the remains of their victims. These are only two of the most dramatic instances of man's destruction of birdlife. Since 1680, the year in which the dodo vanished, seventy-eight bird species have disappeared from the earth. Of these only nine were continental forms, twenty were inhabitants of large islands and forty-nine lived on islands less than 60 square miles in area. In other words, only 11·5 per cent of the vanished birds were of continental origin, the rest being island species.

# Land of the Long White Cloud

It could not have been all that difficult for the hardy Polynesian navigators of the Pacific Ocean to locate and identify the islands later named New Zealand. The halos of clouds encircling the high mountain peaks must have been visible many miles away.

The white-eyes are small New Zealand Passeriformes comprising the family Zosteropidae, with related species in the Ethiopian, Oriental and other parts of the Australasian region. A distinctive feature is the white circle surrounding the eye.

There are many Alpine-type glaciers in New Zealand's mountain regions, the thawing snows of which form rivers and lakes.

In their experience most of the innumerable coral islands and islets in this immense expanse of sea were low-lying, not retaining atmospheric moisture; so one can understand how these huge islands with their towering mountains came to be known to the Polynesian seamen as Aotearoa, the Land of the Long White Cloud.

The modern visitor to New Zealand will find much of the landscape superficially little changed from the time when the predecessors of the Maoris reached these shores about one thousand years ago. He will marvel at the same snow-capped, cloud-encircled mountain peaks and at the tranquil beauty of the glimmering fiords. But he will no longer find these islands harbouring what must have been one of the world's most extraordinary wildlife communities. What particularly astonished the first human settlers were the large numbers of enormous flightless birds known as moas, which played the same ecological role here as, for example, the antelopes and other ungulates elsewhere. These gigantic birds must have winged their way at a very early date to these islands, finding the surroundings ideal, with no herbivorous mammals to contend with them for food. They took over all the habitats normally occupied by phytophages, diversifying to produce numerous subspecies, the largest of which stood 13 feet high and weighed more than 600 lb! Doubtless there were other animals too, such as kakapos (nocturnal parrots) and flightless geese, to surprise the first Polynesian explorers who found this land much to their liking. Their principal source of protein was certainly the huge, inoffensive, flightless moa, together with its eggs and chicks, and this pattern probably lasted for several centuries. But soon after the Maoris arrived the moas had all vanished.

Nowadays, thanks to the temperate climate, the size of the islands and the different types of habitat, ranging from seashore to mountains, New Zealand still boasts a large variety of interesting animals, chiefly introduced by later settlers.

The two principal islands, with a total area slightly exceeding that of Great Britain, are largely mountainous, their coastlines deeply indented. North Island is partially volcanic with geysers, solfataras and fumaroles, all characteristic of this type of terrain and especially numerous in the central volcanic plateau from Lake Taupo to Rotorua. On South Island there are no active volcanoes and the principal geographical feature is the long chain of mountains known as the Southern Alps. These mountains run almost the length of South Island and are eroded by glaciers and rivers which jut far into the forests, while lakes and deep fiords cover ancient glacial valleys. To the east of the Alps there are low foothills and plains.

The climate is temperate and mild, warm in the north, somewhat cooler in the south, with abundant rainfall reaching its maximum intensity in winter, from May to October. The large amounts of rain, chiefly due to the high mountains which attract clouds crossing the Pacific, stimulate luxuriant plant growth.

New Zealand's vegetational cover has been grievously harmed by tree-felling and fires, yet still retains many of its original forest features. In the north-west part of North Island, there are vestiges of a once-impressive forest landscape, with densely packed trees, including the giant kauri (*Agathis*). In the southwest part of South Island, where it is cooler and where rainfall is heavy, there are extensive forests of southern beech (*Nothofagus*) which give way, nearer the centre of the island, to stands of totara and kahikatea, conifers of the genus *Podocarpus*.

Above the forest level there are broad stretches of subalpine vegetation, with flourishing herbaceous plants; but in the much higher alpine areas, especially of South Island, high winds and winter snows prevent the growth of trees or shrubs and confine the plant cover to dwarf forms, mainly grasses and mosses.

Geologists have discovered large quantities of sedimentary rocks in New Zealand. Since these could not have been accumulated by the islands' modest-sized streams and rivers nor by other geological agencies, this is an indication that at one time the islands must have formed part of a continental land mass. But the flora and fauna of New Zealand are remarkably self-contained and individualistic. Many primitive species still exist and a number of important modern animal families are entirely absent; all of which leads to the conclusion that the break with this land mass must have occurred a very long time ago and that the unique nature of the flora and fauna is attributable to a prolonged period of isolation.

The presence in New Zealand of primitive frogs of the genus *Leiopelma* and of that archaic reptile, the tuatara, suggests that approximately two hundred million years ago, during the early Mesozoic era, the two islands were linked to Australia by an isthmus which may have been formed in the Paleozoic era several hundreds of millions of years earlier still. This would explain how invertebrates, primitive frogs and tuataras managed to reach this land. Whatever happened, no such land link existed by the Jurassic period when New Zealand would have been cut off completely, closed to the entry of more highly evolved mammals and reptiles which, had they arrived then, would have brought about the extinction of the older forms.

Mammals, apart from those capable of flying or swimming (there are two bat species endemic to New Zealand and seals are to be seen basking on some beaches), have not been able to cross the strip of ocean separating the islands from other lands. As for birds, they would understandably have had little difficulty in overflying the 1,250 miles or so from Australia. But things were not that simple. A large proportion of New Zealand's birdlife is constituted of ground species that are scarcely able to fly and of a few, such as the kiwis, the takahes (flightless rails) and the now-extinct moas, that are or were incapable of flying altogether. These facts have led some palaeozoologists to conclude that the ancestors of such flightless birds might have emigrated to New Zealand by way of a land corridor existing later than the Jurassic period. But if this were so the route would surely have been open as well to other terrestrial animals. Since no trace of such animal groups has ever been found this theory does not seem very plausible.

The most probable explanation, and that generally accepted, is the following: only flying birds were able to settle in New

The little owls of New Zealand occupy similar habitats and exhibit the same form of nocturnal behaviour as their counterparts in Europe. They feed on insects and small vertebrates.

Zealand after it was completely isolated from the continent. Such birds found a multitude of available ecological niches which they duly occupied. Gradually they acquired terrestrial habits and since they were not threatened by any dangerous natural enemies they eventually lost their powers of flight.

New Zealand therefore merits special attention and study simply because the islands are recognised to be among the oldest regions of our planet and have consequently given rise to a biological community of singular interest and importance. As on most islands, it is not remarkable for an unusual number of plant and animal species (there are relatively few) but for the extraordinary richness of its native flora and fauna.

## Amazing living fossils

For those of us not well versed in zoology—and that applies to the vast majority of people—a frog is a frog and a lizard is a lizard. But the frogs of the genus *Leiopelma* and the tuatara of the genus *Sphenodon* are something special. Although not unusually distinctive in appearance, they are recognised as genuine living fossils. From the scientific point of view it is as if there were still some dinosaurs at large.

There are three species of these primitive frogs, *Leiopelma archeyi*, *Leiopelma hochstetteri* and *Leiopelma hamiltoni*, named after the people who discovered them. Together with another primitive North American frog of the genus *Ascaphus* they belong to the family Leiopelmidae, whose principal characteristics are the possession of so-called amphicoelous vertebrae (concave on both sides and hence different from those of other amphibians but similar to the teleostean fishes), a caudal muscle structure—although no tail in the adults—and a male copulatory organ. These anatomical features indicate that the animals are very primitive and consequently exceedingly rare.

The little New Zealand frogs, measuring little more than 2 inches in length, are usually found in or near mountain rivers and streams, depositing their eggs underneath stones, in the hollows of dead trees or sometimes in tunnels that have been dug in the mud by burrowing insects.

Unlike other frogs these species are not very adept swimmers so that the major part of their time is spent on dry land. Furthermore, the process of metamorphosis is highly specialised, for everything happens inside the egg. Thus in the forty-one-day interval between egg laying and hatching the unicellular fertilised egg is transformed into a tadpole, thus completing the early stages of development. When the youngster is hatched it already looks like an adult frog except for its long tail which helps the animal to breathe through the skin (thanks to numerous capillaries) until the lungs are fully developed. As it takes on fully adult characters the tail disappears, but not the musculature, which is retained throughout life.

The tuatara (*Sphenodon punctatus*) possesses even more archaic features and is the most striking living proof of the long geographical isolation of these regions. Of the ancient order of Rhynchocephalia, reptiles that vanished during the Jurassic

The little frogs of the genus *Leiopelma* do not appear to be very different from other species; but they are in fact among the most primitive animals on earth, special anatomical features being vertebrae that are concave on both sides and a caudal muscle structure, despite the fact that the adults have no tail.

period, this is the only genus and in fact the only species of that genus to have survived. To study the biology and behaviour of the tuatara is to step back about two hundred million years into prehistory and the great age of reptiles. This inoffensive, lizard-like inhabitant of the islands off the New Zealand coasts is a direct descendant of reptiles living at that time.

Of all the anatomical characteristics of the tuatara, the most remarkable (common to many fossil vertebrates) is the possession of a third eye known as the pineal eye, lying beneath an opening in the skull on top of the head, much more rudimentary than the two real eyes and with faint traces of a retina and lens. This organ, lying under the skin and resting on the pineal gland or epiphysis, is also found, although in an even more vestigeal form, in many living lizards, but in none is it as large or as recognisable as a true eye as in the tuatara.

The function of the pineal eye is not known. In lizards it would seem to act as a kind of sun-meter, regulating the time the animal exposes itself to the sun, a very important consideration for a cold-blooded animal. It is possible that it has a similar function in the case of the tuatara, although being so well developed it might provide a better receptor, with physiological repercussions that remain undetermined. In any event it would certainly seem to be indirectly related to temperature regulation.

The tuatara, like the primitive frogs, also has amphicoelous

The tuatara is the only surviving representative of an order of reptiles which lived some two hundred million years ago and then died out except for this one species. Its presence on some of the islands around New Zealand proves that the country was separated from continental land masses a long time ago, certainly before the Jurassic period.

vertebrae, further proof of its primitive origins; but unlike other reptiles it lacks a male copulatory organ.

The tuatara's life pattern is different from that of later reptiles. In the first place its metabolism is very slow, the slowest in fact of all living vertebrates. The animal can go without breathing for an entire hour. Secondly it has a long life span (sometimes more than a hundred years) and requires very little heat, so much so that it can remain active at temperatures as low as 14°C (57°F) or even 11°C (52°F). No reptile has a stronger resistance to cold, yet despite this the sudden lowering of temperature in autumn compels the tuatara to find a convenient place of refuge where it remains in a state of lethargy during the winter, re-emerging in spring. In this respect the tuatara enters into a surprising association—a matter of tolerance rather than symbiosis—with local species of shearwaters and petrels which nest in tunnels along the seashore. When the time comes to retire the tuatara simply crawls into a bird's nest. The legitimate occupant seems unperturbed, though not unaffected, by the intruder.

Study of the habits of these so-different animals shows that it is the reptile which derives the greater benefit from this apparently harmonious relationship for it finds a well prepared refuge in addition to scraps of food. It is very probable that the burrowing birds play host to the tuatara even outside the period of hibernation. But things may not be quite as peaceful as they might appear, for according to some observers, although its normal diet consists of insects, earthworms and snails, the tuatara is likely to consume the eggs or nestlings of its host, although it seems not to molest the adults.

## The moa: giant bird of the past

Seven or eight centuries ago flocks of enormous moas roamed the plains of New Zealand as freely as herds of zebras and antelopes in modern Africa. Prior to Charles Darwin's celebrated voyage in the *Beagle* (1831-36), in the course of which he visited New Zealand, nobody in the West had heard of these birds and it was he who first aroused the curiosity of other European naturalists in this strange and evidently extinct species. Darwin was particularly interested to discover the reasons for many endemic species of birds and insects having lost their powers of flight and his conclusion was that the absence of wings was in itself a positive aid to survival in an island environment where the extremely violent local winds might well have spelt disaster for smaller winged species.

Among the world's surviving animals only the giraffe is taller and the African elephant of greater weight than the largest recorded moas. Such huge dimensions initially puzzled zoologists. Sir Richard Owen, the first scientist actually to examine a bone of a moa automatically assumed that he was handling the femur of a cow or horse but further study convinced him that it belonged to a running bird. Later, in 1843, Owen, now furnished with a great deal more material, revealed that several species were involved but that none of them had ever possessed the bony wing structure of modern birds. Ornithologists think that there

*Facing page:* The kiwis are strange flightless and tailless birds with nocturnal habits. Unlike other birds their nostrils are situated at the tip of the long curved bill and it is by their sense of smell that they locate the insect larvae and earthworms on which they feed in the winter months.

Little spotted kiwi
(*Apteryx oweni*)

Common kiwi
(*Apteryx australis*)

The little spotted kiwi, smaller of the two spotted species, both found on the South Island. The common kiwi lives on South Island, North Island and Stewart Island, each island having a distinct subspecies.

*Facing page*: Kiwis move with astonishing speed and assurance through the undergrowth, seeking their food by night. The female lays one or two eggs each weighing about one pound, unusually large and heavy for a bird of this modest size.

were between twenty and twenty-seven species.

The moa was evidently a peaceful bird, basically herbivorous, leading a type of life comparable to that of certain ungulates such as the giraffe. In addition to grass it probably ate insects, molluscs, crustaceans and fishes; and judging by its size and weight it must have consumed as much food daily as a cow.

Profiting from the work of Owen and other specialists, ornithologists have speculated as to why these birds vanished. At the beginning of the 20th century New Zealand was frequently visited by zoologists who closely questioned the Maoris and were regaled with graphic tales of moa hunts and feasts, such stories being embroidered with highly coloured descriptions of the birds' alleged habits. Although there was hardly any vestige of truth in such tales (for no Maori ever admitted having seen a live moa) the fact remains that their folklore was full of references to this gigantic bird and that the stories must have been handed down by word-of-mouth from one generation to another.

The early Maoris certainly encountered moas, so the belief that the entire group must have been extinct before the first Maoris arrived is incorrect. These bold seamen are thought to have set out from Polynesia (or perhaps South America) around 1350, three centuries before any Europeans touched the islands. But recent discoveries show that there were earlier arrivals still. Remains of sites and artefacts indicate that these immigrants, certainly from Polynesia, did hunt and eat moas and would often bury their dead together with moa bones and egg shells. When the Maoris reached the Land of the Long White Cloud they did not find them uninhabited and they must have fought and conquered their predecessors who had doubtless exterminated several species of moa. The moa bones and feathers, together with tools and weapons fashioned from parts of their bodies, might have been collected by the Maoris as war trophies.

It is likely, nevertheless, that some moas, notably smaller species in South Island, were still around when the Maoris arrived; but these too would have been speedily eliminated by the time the Europeans stepped ashore in the 18th century.

Some of these giant birds are still being uncovered. The oldest fossils embedded in a layer of sedimentary rock date back about two million years. The draining of a swamp also turned up skeletal remains of birds that had evidently drowned. Radioactive carbon tests established the approximate period when such birds lived for the stomach contents of one were composed of material ingested in 1290, shortly before the first wave of Maoris came.

## The enigmatic kiwis

The discovery in New Zealand of another flightless bird, about the size of a large chicken (weighing between 2 and 10 lb) and apparently related, albeit distantly, to ostriches, rheas, emus and cassowaries, has led experts to the conclusion that the ratites are a polymorphic group. The birds comprising this group probably did not have a common ancestor and would have appeared simultaneously in different parts of the world.

Geographical distribution of kiwis.

**COMMON KIWI**
(*Apteryx australis*)

Class: Aves
Order: Apterygiformes
Family: Apterygidae
Total length: 18-22 inches (45-55 cm)
Weight: male 4½ lb (2 kg)
female 6½ lb (3 kg)
Diet: invertebrates, mainly earthworms, leaves and fruit, the vegetation being eaten mainly in the dry season.
Number of eggs: 1-2
Incubation: 75-80 days

Plumage consists of long, loose, drooping, brownish grey feathers. Nostrils situated at tip of upper mandible, the bill measuring 4-8 inches. Rudimentary wings; no tail. Very powerful feet with three strong forward-pointing toes and one shorter toe at the rear. Two other species are the great spotted kiwi (*Apteryx haasti*) and the little spotted kiwi (*Apteryx oweni*).

The common kiwi (*Apteryx australis*), whose Maori name is derived from the strange call of the male bird, has a number of subspecies. It is an inhabitant both of North and South Islands as well as Stewart Island. Two other species are recognised by some ornithologists in the western and southern parts of South Island, namely the great spotted or Haast's kiwi (*Apteryx haasti*) and the little spotted or pygmy kiwi (*Apteryx oweni*), also with several subspecies.

The kiwi, New Zealand's national emblem, is a curious-looking bird with a spindle-shaped body covered with loose feathers that grow like hairs over rudimentary wings. The bird has no tail. The legs are short and the four-toed feet very strong. The head terminates in a long, flexible bill, the lower mandible being somewhat shorter than the upper. The nostrils are situated at the tip of the upper mandible, this being an exceptional feature for among the vast majority of birds the nostrils are placed at the base of the upper mandible. The eyes are small and the ear opening large.

Given that the kiwi is a nocturnal bird and that the eye structure appears to preclude good night vision, it has always been assumed that it manages to find its direction in the darkness by means of hearing and smell. Recent experiments by Professor B. M. Wenzel of the University of California have shown in fact that the kiwi does indeed have an extremely highly developed sense of smell. In his observations in New Zealand Professor Wenzel buried under the floor of an aviary containing several kiwis various food substances in aluminium tubes. Switching the contents of the tubes and testing different birds, he proved beyond any doubt that the kiwis can easily locate, by smell alone, a variety of foods buried almost a foot below ground.

Kiwis inhabit forests with dense vegetation (ferns, kauris, etc), preferably in areas where the soil is wet. They are excellent runners, moving with a strange, rolling gait, but since they are normally active only at night they are not easy to observe and consequently very little is known of their behaviour in the wild. On the other hand their anatomy and morphology have been intensively studied.

The food habits of these birds assume two distinct patterns, according to the time of year. In the rainy season they feed principally on small invertebrates, especially earthworms, which they evidently locate by scent and probe out with their sharp bill, as well as insects lodged in tree bark. It is possible that they also use their acute hearing to locate their prey much in the manner of woodcocks, which locate worms underground by the sounds they make. In summer, when the ground is dry, the kiwis turn herbivorous and feed on leaves and fallen fruit.

Hardly anything is known about the social behaviour, territorial habits, moulting procedures or longevity of kiwis in the wild; but it has been verified that once a pair have mated it is the smaller male who takes charge of the incubation. The nest, which is generally situated between the buttress roots of a tree, is a simple depression in which the female lays one or two eggs, depending on the species. A subspecies from North Island, *Apteryx australis mantelli,* lays two white eggs each weighing

The weka, a flightless member of the rail family, spends much of its time searching for food which consists in the main of insects.

about 1 lb. This is extremely large in relation to the size of the bird's body, equivalent to perhaps one-quarter of the female's weight. After about 75-80 days' incubation the chick hatches, hiding in the nest and taking no food for six days, then feeding in company with its father.

Before the Europeans arrived kiwis were abundant in New Zealand. The Maoris regarded them as food delicacies but killed no more than were necessary to satisfy their needs. White settlers, during their systematic programme of deforestation, brought about a sharp decline in the kiwi population. This was aggravated by their dogs which were more than a match for the flightless birds. Yet despite the changes in their habitats and the introduction of predators such as dogs, cats and stoats, the kiwi, although much reduced in numbers, is not considered an imminently threatened species.

The takahe is a large flightless rail believed to be extinct until it was 'rediscovered' on South Island in 1948. It is a protected species and there are thought to be about three hundred surviving birds.

# The flightless rails

Of all the groups of birds inhabiting the world's oceanic islands, the Rallidae, comprising rails, gallinules and coots, appear to have been most affected by natural selection in their efforts to adapt to their environment.

It is not surprising that the members of this family should have lost the faculty of flight, given the fact that they are essentially swimming or wading birds which, in the event of danger, either use their long legs to run away or conceal themselves among water plants. Doubtless, a long time ago, certain Rallidae of South-east Asia were blown off-course during their seasonal migrations, reaching islands where they found food in plenty and no predators. Since they were already largely terrestrial in habit and now encountered ideal conditions which merely reinforced such tendencies, they not only lost their flying powers but increased in size, well able to cope with all emergencies thanks to their long legs and extremely strong beaks.

Palaeontologists have recovered various forms of extinct Rallidae from fossil beds in New Zealand, the skeletons of which do not appear to resemble those of flying birds. Today three endemic species belonging to this family still live in the islands, one of which, the takahe (*Notornis mantelli*), a large flightless gallinule, is in grave danger of extinction. This bird is an inhabitant of the Southern Alps, frequenting patches of tussocky grass and undergrowth. Blue, with a red frontal shield and powerful bill, it feeds on plants and seeds—a winged counterpart of herbivorous mammals of other regions.

There are believed to be only about three hundred takahes left and their plight is not improved by the fact that their rate of reproduction is low, each female laying one to four eggs, some of which fail to hatch.

The second representative of the Rallidae still surviving in New Zealand is a much smaller brown bird known as the wood rail or weka (*Gallirallus australis*) with several subspecies. This completely flightless rail is a predator, scouring the undergrowth for small rabbits, mice, lizards, other birds' nests and insects. It is frequently seen too on river banks, hunting for fishes, snails and small crustaceans.

The most common endemic rail is the pukeko (*Porphyrio melanotus*) which like the takahe has a massive bill and red frontal shield. Unlike the other two endemic species, the pukeko can fly quite strongly over short distances. It lives mainly in swampy areas, but may be seen feeding in open grassland nearby. It has a parrot-like habit of sometimes holding its food firmly in one foot.

On the island of New Caledonia, in the South Pacific east of Australia, lives another curious greyish-white bird, about the size of a large hen, which is barely able to fly. The kagu (*Rhynochetos jubatus*) was once very abundant and is today on the brink of extinction. The sole member of the family Rhynochetidae, it appears to be related to the sun-bitterns of South America, with similar long legs for wading, a bright orange beak and a crest of long grey feathers on its head. The body feathers are loose.

Ponds and swamps are favoured habitats of the pukeko (*Porphyrio melanotus*), distinctive for its powerful red bill and frontal shield.

*Facing page:* The kakapo or owl parrot, as its second name suggests, is a curious bird classified as a member of the parrot family but with the nocturnal habits and facial features of an owl. It is a silent bird with nondescript greenish plumage and is incapable of true flight.

---

**KAKAPO OR OWL PARROT**
(*Strigops habroptilus*)

Class: Aves
Order: Psittaciformes
Family: Psittacidae
Total length: 23-26 inches (58-65 cm)
Diet: basically grass and leaves, from which it extracts the juices
Number of eggs: 2-4

Large flightless parrot with nocturnal habits, running and climbing expertly and sometimes making short glides. Plumage green with black, grey and brown streaks. Owl-like facial discs; whitish bill; fairly long tail; feet fairly short and strong.

# The kakapo, a nocturnal parrot

Ornithologists were for long in disagreement about the origin of the kakapo (*Strigops habroptilus*). Some argued that it was without doubt a member of the parrot family (Psittacidae); others judged it to be a link between the parrots and the nocturnal birds of prey (Strigidae). The former are of course generally diurnal by habit, brightly coloured and noisy in flight. The kakapo, on the other hand, is a nocturnal ground bird with fairly drab plumage, incapable of true flight, and silent. Although nowadays accepted as one of the Psittacidae, its special status in the family is underlined by the alternative name it has been given, the owl parrot.

Although the kakapo's wings are well developed the musculature is weak. Just the same, the bird will sometimes glide on thermal currents from valley to valley. On very rare occasions it may even lift itself into the air to avoid an obstacle or to reach a selected perch. Its greenish plumage is irregularly streaked with blacks, greys and browns. This comparatively unspectacular coloration undoubtedly has the effect of camouflaging the bird during the daytime when concealed among rocks or in the undergrowth, so that it can be relatively secure from the usual diurnal raptors.

The feathers of the face and head have also undergone modifications in comparison with those of related species. Thus these birds possess facial discs and the vibrissae at the base of the beak are elongated, characteristics that are untypical of parrots but found in many owls.

In the 19th century the kakapos were the most numerous of all New Zealand's ground birds; but since 1900 the populations have declined in alarming fashion, to the point that the species is well-nigh extinct (it is thought to have vanished already from North Island). The two factors principally responsible for this sad situation are, firstly, the importation of stoats and, secondly, the introduction of herds of deer which have completely transformed the environment.

The nocturnal habits of the birds make them particularly vulnerable to the predatory activities of the night-hunting stoats. Sometimes, in order to move around, kakapos will open paths through the low vegetation by pecking at roots and plants, affording them some protection from enemies; but these are invariably trampled underfoot by roaming deer. Such intrusions by alien species have consequently led to the disappearance of kakapos in most areas, so that one can generalise by saying that the birds now occupy, to a steadily decreasing extent, only those regions where deer have been comparatively recently introduced.

Since 1958 various nature conservation groups have organised fifty-five expeditions to the Southern Alps with a view to studying the behaviour of the kakapo. Among them they managed to sight the birds in only eight localities, confirming that they are indeed exceedingly rare on South Island and explaining why so little is known of this extraordinary species.

In the IUCN Red Data Book of threatened animal species

The yellow-crowned parrakeet, one of the four endemic New Zealand species generally nests and roosts in trees but in coastal areas often does so in rock fissures.

much space is devoted to the tireless work being done by New Zealand conservation authorities to discover the true state of affairs. All the information so far provided is profoundly disturbing. In 1960 the total number of kakapos was estimated at less than two hundred individuals; but by 1970 this figure had gone down by half. Such statistics show indisputably that the future of the kakapo is indeed black, taking into account the fact that all attempts to reintroduce the species to previous haunts with birds raised in captivity have failed.

Adult kakapos are known to feed principally on vegetation, although they do occasionally eat small lizards. Like many other members of the parrot family they tend to waste a good deal of food, and their presence in a particular place is often betrayed by little heaps of broken-off twigs and leaves which they have left untouched. During the day the birds sleep in hollows at the foot of trees or in rock clefts.

The birds do not breed regularly every year but usually in alternate years. They build their nests in similar sites to those used as dormitories and the female lays two to four white eggs. She incubates them alone and raises her brood without the assistance of the male.

## New Zealand's parrakeets

The origin of the four species of parrakeets still to be found in New Zealand, all belonging to the genus *Cyanoramphus,* may perhaps be traced to a common ancestor from which also stemmed the Australian parrakeets. All of them have greenish upper parts, yellow underparts and bluish wing coverts, and both tail and feet are comparatively long.

The red-crowned parrakeet (*Cyanoramphus novaezelandiae*) has the broadest range of distribution of these local species for it is an inhabitant not only of New Zealand but also of many of the surrounding islands. As its name suggests, the upper part of its head is bright red, easily distinguishing it from related species. The yellow-crowned parrakeet (*Cyanoramphus auriceps*), although more numerous, has a much more restricted range. Apart from the yellow coloration of the top of its head it differs from the red-crowned species in being rather smaller.

Both these parrakeets live in forests and it is certain that in days gone by they enjoyed a far wider range. Today, mainly as a consequence of deforestation, they have disappeared from many places where the remaining patches of vegetation have proved inadequate to provide them with enough sustaining food. The red-crowned parrakeet is now very rare but the yellow-crowned parrakeet has recently undergone a moderate expansion in the larger forest tracts. There seems to be some overlap in food preferences, both species eating a variety of vegetable matter including buds, leaves, fruit and seeds. There is one major difference, however, as is usual where two closely related species occupy the same area; red-crowned parrakeets descend to the ground much more frequently to feed on ground plants.

While these two species were evolving, two other much more specialised parrakeets also made their appearance. One is the

orange-fronted parrakeet (*Cyanoramphus malherbi*) which is found only in high mountain districts, the other is the Antipodes Island parrakeet (*Cyanoramphus unicolor*). The latter lives only in the Antipodes, a group of small uninhabited islands lying to the south-east of New Zealand.

Both the Antipodes Island and red-crowned parrakeets occur on the Antipodes, the former having arrived much earlier than the latter. The Antipodes Island parrakeet displays all the typical features of an island bird, being markedly larger and more sparsely feathered than its continental relatives. In its habitat the plant cover is poor, consisting mainly of tussock grasses and some low shrubs which provide food in the form of buds,

The red-crowned parrakeet is a handsome little bird with a red head, crimson eyes and green plumage. Its main food is a wide range of vegetable matter, often from the ground, and it has suffered heavily from deforestation and encroachment of man and his introduced animals.

Takahe (*Notornis mantelli*)
Kagu (*Rhynochetos jubatus*)
Weka (*Gallirallus australis*)
Kaka (*Nestor meridionalis*)
Yellow-crowned parrakeet (*Cyanoramphus auriceps*)

leaves, fruit and seeds. These lonely islands are also the nesting sites of immense penguin colonies for some eight months of the year, and these provide a seasonal supply of broken eggs, skins and carcasses of dead penguins, dropped by the flocks of giant petrels (*Macronectes giganteus*). The Antipodes Island parrakeet makes considerable use of this supplementary food, and it may be this divergence which enables it to share the island with the red-crowned parrakeet.

## A carnivorous parrot: the kea

Of all the Psittacidae the most astonishing examples of adaptations to a varied diet are exhibited by the kea (*Nestor notabilis*) and the kaka (*Nestor meridionalis*). Both these endemic parrot species have lived for so many centuries in New Zealand that it is extremely difficult to discover those characteristics that link them with the rest of their huge family.

In many ways these birds behave in a similar manner to the cockatoos of Australia for in their case too it is evident that they show little fear of man and that they rapidly become accustomed to living alongside him.

Besides possessing rounded wings the kea exhibits several other typical features such as a disproportionately long upper mandible, which is sharply curved, and brilliant orange hues on the undersides of the wings.

The kea is an excellent flier and is an inhabitant of mountain valleys rather than forests, nesting in rock fissures and on cliff ledges. In hard winters it descends to the plains and sometimes to the seashore.

The name kea is also of Maori origin and derived from the call of the bird in flight. This species prefers a mountain habitat and is often seen in the Southern Alps, occupying the last vestiges of virgin tropical forest; but it is frequently encountered much higher where woodland gives way to rocky terrain with a rampant cover of low shrubs.

In winter, snowstorms drive the keas, particularly young birds, down to the plains, sometimes as far as the seashore. Adults tend to be more sedentary in habit than their offspring and only abandon their territory in exceptional circumstances when food is otherwise impossible to obtain. In other words the parents sacrifice living space in favour of their young ones.

Of all the New Zealand birds the keas are the most persecuted, doubtless because of the sinister reputation they have among stock-breeders who allege that they decimate their flocks. Although the harm they do is greatly exaggerated there is no denying the fact that keas do feed occasionally on the kidney fat of sheep, attacking live animals as well as alighting on carcases. Since sheep constitute one of the principal natural resources of the country it is not surprising that the misdeeds of keas receive a great deal of adverse publicity. Yet the ornithologist J. R. Jackson, who examined the dead bodies of some two thousand sheep in different mountain localities, affirmed that in not a single case was he able to state definitely that death had resulted

The kea, with its inconspicuous plumage and enormous curved bill, is one of the strangest members of the parrot family, having adapted to its environment by adding flesh (or more specifically fat) to its normal vegetarian diet. Because of its alleged attacks on sheep, farmers have subjected the species to intense persecution.

from the attack of a kea. The bird's reputation as a sheep-killer has become a part of New Zealand folklore, so that it is virtually impossible to sort fact from legend. Many sheep obviously die every year from natural or accidental causes and in the vast majority of cases the parrots have simply been seen pecking at carcases. Farmers have sometimes set traps for the birds by baiting the bodies with poison or capture kea chicks in the nest and use them to entice other birds with their shrill cries. The theory has also been advanced that only certain keas attack livestock and this may well be true. In any event the authorities show a somewhat ambivalent attitude towards the birds. In some areas they are even afforded official protection.

Keas are in fact omnivorous. In addition to carrion they feed on tender shoots, fruit and nectar extracted with their bristly tongue. But they play no part in plant pollination for they simply tear up the petals to get at the sweet liquid inside the corollas. This behaviour seems to indicate the primitive nature of the species.

Spring and summer are the seasons for rearing the young. Nests are built in rocky fissures where the female lays two to four whitish eggs. The male does not assist in their incubation but attends to the food requirements of his partner and offspring. The latter are ready to fly between the ages of thirteen and fourteen weeks. The males may be polygamous. The females do not necessarily lay every year, and on average each rears one young every second year. As a consequence, there needs to be a life expectancy of at least six years to maintain their numbers.

The species, much persecuted by man and very vulnerable to winter cold, is additionally victimised by stoats and rats. For those reasons the kea does not usually live long in the wild. Only in national parks, where the species is properly protected, can it expect a reasonable longevity. One individual ringed in a reserve in 1956 was found alive twelve years later. Closely related to the kea, but without the latter's evil reputation, is the kaka, widely distributed through all of New Zealand's forests. The kaka's plumage is much more vivid and handsome than that of the kea and the upper mandible shorter. The bird frequently nests in the trunk of a tree, boring its own hollow. Although in some areas it is confused with the kea and therefore hunted, the kaka is generally protected. It is of shy disposition and easily domesticated. In times of famine the Maoris fed on the flesh of this bird and arrayed themselves with kaka feathers which were also used as monetary units.

## Privacy invaded

Because of its isolated geographical position in the southern hemisphere New Zealand, although subjected to successive waves of migration over the centuries, was one of the last regions of the Pacific to be settled. Once the explorers and colonisers reached its shores, however, the cumulative effect of their activities proved to be profound and, in many respects, highly destructive. It is no exaggeration to say that these islands are perhaps the most shameful testament to man's interference with

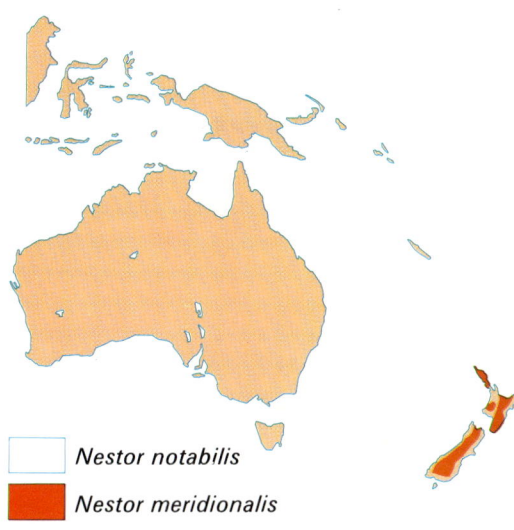

Geographical distribution of the kea (*Nestor notabilis*) and the kaka (*Nestor meridionalis*).

---

**KEA**
(*Nestor notabilis*)

Class: Aves
Order: Psittaciformes
Family: Psittacidae
Total length: 17½-19 inches (44-48 cm)
Diet: once exclusively vegetarian (fruit, flowers, nectar, leaves, etc.) but, with the appearance of domestic sheep, some birds have come to feed on carrion and are said also to attack live animals.
Number of eggs: 2-4
Incubation: 22-29 days

Parrot, approximately the size of a large crow, with an extraordinarily long, powerful bill, the upper mandible sometimes more than 4 inches. Upper parts olive-green, underparts yellower with reddish veining. Lower wing coverts orange-red, feet yellowish-brown. Closely related to kaka (*Nestor meridionalis*) which does not appear to have developed carnivorous habits.

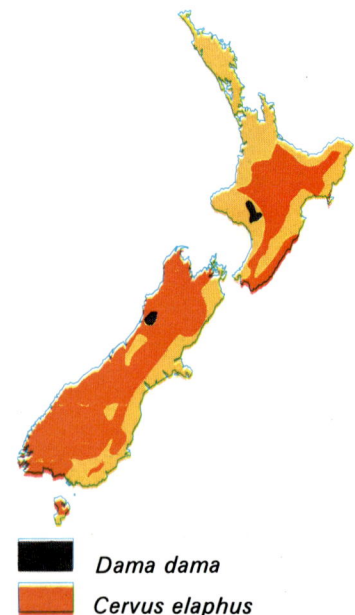

**Dama dama**
**Cervus elaphus**

Among the many mammals introduced by man into New Zealand, the red deer (*Cervus elaphus*) has, as the above diagram indicates, attained the widest distribution. The fallow deer (*Dama dama*) has a very much more restricted range. But both species have helped to transform the original vegetational cover, with damaging effects on endemic forms of wildlife.

a natural environment and its inhabitants.

The first Polynesian navigators are thought to have arrived about one thousand years ago and at that time the islands would have been virgin territory, natural sanctuaries of plant and animal life completely different from anything that existed at that time in other parts of the known world.

It was the earliest arrivals who unwittingly upset the delicate balance of nature by hunting the giant moas—easy enough victims—for their flesh and feathers. When the Maoris arrived, some three hundred years later, only a few scattered flocks of the huge flightless birds were left; and these too were rapidly exterminated, together with the poorly armed primitive tribes that had originally hunted them.

The Maoris brought with them to New Zealand two types of animal—the dog and the Polynesian rat (*Rattus exulans*). Surprisingly, the latter does not appear to have had disastrous effects on the local forms of wildlife.

In 1769, in the course of his first voyage, Captain Cook landed in New Zealand. His subsequent visits in 1774 and 1779 and the ultimate European settlement of the islands led to the progressive importation of alien species. Cook's men themselves planted cabbages, beetroots and other vegetables and left behind them goats, sheep and pigs, some of which survived and bred. This was the vanguard of a massive invasion in which 168 species of birds and mammals were introduced, only 65 of which managed to adapt successfully to their new surroundings.

Among the animals that did succeed in settling down in their new homeland were various forms of deer (red, fallow, axis, sika and sambar), hares, rabbits, elks, chamois, horses, cattle and certain species of wallaby. The policy of introducing such a large number of plant-eating animals led inevitably to an excess of herbivores which, because they were not faced with a proportionate number of predators who would have controlled their populations, proved well able to cope with the relatively few small carnivores introduced at about the same time. The only available predators were stoats, weasels, dogs and domestic cats. Not unnaturally these were incapable of playing any kind of significant role in controlling the growth of the phytophage communities.

Precisely the same happened with the birds. Enormous flocks of pheasants, partridges, mallard, black swans, Canada geese, thrushes, blackbirds, starlings, larks, linnets, goldfinches, sparrows and so forth were imported at random, without any thought for the possible consequences of their not being controlled by specific predators. The worst enemies they had to face were native falcons (*Falco novaeseelandiae*) but these could never have been very abundant. The harrier hawk (*Circus approximans*) also takes small birds occasionally.

In the case of imported plants the end result was not so disastrous. Of the 603 imported species only 48 were found to threaten the future of existing native species.

The repercussions of this indiscriminate policy of importing alien species of animals—mammals and birds in particular—were felt almost immediately. Taking advantage of an ideal environ-

ment where so many contrasted habitats were freely available and where there was little if any danger from natural enemies, the communities of introduced animals underwent a population explosion of unbelievable proportions, the consequence of which was a virtual ecological cataclysm. The herbivores, in fact, were free to wander everywhere, ruining the soil and devastating vegetation from ground to treetop level. Endemic species, driven from their traditional haunts by famine, were eventually wiped out for they were unable to compete with the more highly evolved intruders.

Man alone, through his recklessness and stupidity, has been responsible for the destruction of one of the world's last remaining Gardens of Eden. The destruction of this island paradise did not come about as a result of deliberate cruelty or greed but by the simple transgressions of nature's immutable laws. It would take a complete reversal of policy and centuries of enlightened endeavour to restore the natural wealth of New Zealand's flora and fauna. As the French zoologist Jean Dorst has pointed out, it is as if New Zealand had been earmarked to show precisely how the balance of nature can be destroyed by human ignorance and folly.

This aerial photograph shows how the virgin forests of New Zealand have been hacked or burned down to make way for pasture land. In many areas deforestation has proceeded so recklessly that the exposed, unprotected terrain has been unsuitable for plant growth and been transformed into wasteland. Fortunately there are still large areas of undisturbed forest and woodland on both islands where endemic species of plants and animals can still flourish.

# CHAPTER 104

# Madagascar, isle of wondrous plants and animals

In the Indian Ocean, so close to the African east coast that it seems to be a piece of land wrenched from the Dark Continent, is a huge island. With an area of more than 225,000 square miles it is the fourth largest in the world. This is Madagascar which, together with tiny adjacent islands, forms the Malagasy Republic.

In the extreme north, close to the equator, the climate is humid and tropical, but a thousand miles away in the south it is temperate and relatively dry. Running like a backbone down the centre of the island is a chain of mountains, averaging 5,000-6,000 feet in height but with some peaks towering above 9,000 feet. On the east coast frequent heavy rainfall stimulates the growth of luxuriant vegetation, characteristic of the tropical rain forest; and along this coast too there are mangrove swamps, marshes and lagoons. On the opposite side of the mountains the climate, slightly drier, has favoured the development of open deciduous forest which, in many places, because of human activity, has degenerated into plain and savannah, sparsely covered with bushes and trees. In the very driest southern regions the landscape takes on a sub-desert appearance, with cactuses, euphorbias and baobabs. Nowhere else in the world are these species all found together.

The plants and animals of Madagascar are truly astonishing but naturalists who have not actually visited the island are strangely ill-informed about them. Sherwin Carlquist hazards the guess that if the average zoologist were asked to name the typical animals of Madagascar he would mention the lemur family and then pause to reflect; nor would a botanist who has not made a specialised study of the island's flora be able to name more than two or three genera of plants. Although plants are

*Facing page:* Madagascar, fourth largest island in the world, still harbours a fascinating variety of strange plants and animals. The latter include many reptiles such as this flat-tailed gecko, a masterpiece of natural camouflage.

On Madagascar, where the landscape ranges from tropical forest to steppe and thorny desert, many beautiful flowering plants grow, including euphorbias such as (*Poinsettia madagascariensis*), pictured here.

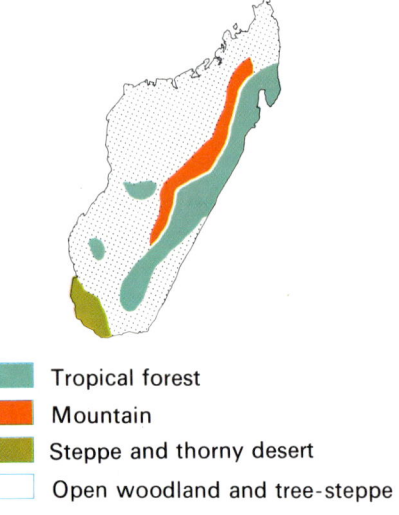

- Tropical forest
- Mountain
- Steppe and thorny desert
- Open woodland and tree-steppe

Original vegetational zones of Madagascar.

not the central theme of this work, it is worth mentioning the spectacular genus *Bubbia*, first discovered in 1957. Its nearest relatives are to be found in distant New Guinea.

## Oceanic or continental island?

Madagascar is separated from Africa by the Mozambique Channel which at its narrowest point is less than 250 miles wide. But has this arm of the sea dividing island from continent always been there? Can one determine with any certainty whether this is an oceanic or continental island?

From the geological point of view the channel is sufficiently deep and the types of terrain on the African mainland exhibit such marked differences that it is safe to say that Madagascar has not been joined to the continent in comparatively recent times. But this only proves that Madagascar has been an island since antiquity, not that it never formed part of a single vast land mass.

Since geology cannot provide the answer to the question one has to turn to the evolutionary evidence of the island's plants and animals. Perhaps the distribution of flora and fauna can solve the problem of Madagascar's origins.

The island is certainly poor in vertebrates and the majority of African species are unrepresented here. Among Madagascar's mammals it is considered probable that the more modern rodents, shrews and bushpigs have all been introduced by man comparatively recently; but one cannot entirely dismiss the possibility that some may have arrived by other means, for many animals apparently incapable of crossing expanses of water have shown themselves to be amazingly adept at swimming. It is not impossible that pigs might have reached the island in such a way.

As for civets and primates, they are recognised as characteristic island colonisers so that it is not surprising to find them in fair numbers here.

The insect-eating Tenrecidae and the endemic rodents of the island are two typical examples of the development of groups after having been separated from others of their kind. In Madagascar they have almost no competition for living space in their chosen habitats. As one might expect too, there are vast numbers of bats on the island, some of African, some of Oriental, origin—understandable, given the mobility of these hardy flying mammals.

Closer study of the island's birds, reptiles, amphibians and fishes fails to throw much more light on the matter. But the discovery of fossil remains of dinosaurs and the fact that many of the animals in Madagascar are of fairly ancient stock leads to the conclusion that if the island was indeed joined to Africa it must have been many millions of years ago that they became finally separated, probably before the beginning of the Tertiary period.

Certain plants, birds and amphibians show links with similar forms in tropical Asia. It is this which has provided support for a theory that at one time Africa and Asia formed one huge land mass known as Lemuria, of which Madagascar was a part. But this remains an assumption, without any really strong evidence;

Three-quarters of Madagascar's land surface is today given over to agriculture. This plantation with its neat paths has sprung up on the site of what was once luxuriant jungle.

if Lemuria ever did exist it must have been at least two hundred and twenty-five million years ago, before the Triassic period.

To sum up, Madagascar is probably a continental island where millions of years of isolation have given rise to strange examples of evolution and adaptation similar to those that are to be encountered on oceanic islands. For that reason, as Carlquist has pointed out, all the characteristic phenomena of island life—adaptive radiation, living fossils, endemic species, gigantism, terrestrial birds, and animals teetering on the verge of extinction—are to be found on Madagascar.

## Fishes and amphibians

Of the 48 families of freshwater fishes inhabiting the lakes, streams and rivers of Africa, only 23 are found in Madagascar. Strictly speaking, there is not a single genuine freshwater species here for all are descended from marine fishes that have adapted to life in slightly salty water.

Although there are some 155 species of amphibians they do not show much variation and there is a high degree of endemism. There are no salamanders or newts, nor any toads and tree frogs. The family Rhacophoridae, however, is especially well represented and the frogs of the genus *Rhacophorus* are the most numerous. This genus has undergone a speciation explosion on Madagascar and there are at least 44 species.

Two genera of the family are also represented in Africa, one in Asia, and four are endemic. The Rhacophoridae look somewhat like miniature tree frogs but there are in fact significant differences in the structure of the skeleton.

## The out-of-place iguana

Reptiles are very abundant on Madagascar and the iguanas, in particular, are of especial zoogeographical interest.

Iguanas are of course characteristic saurians of the New World and their counterparts in Eurasia, Africa and the Oriental region are the agamids. It is astonishing enough that iguanas of the genus *Brachylophus* should have reached the remote islands of Fiji and Tonga, but even more amazing that two genera, *Chalarodon* and *Oplurus* (the latter represented by six species) should be living on Madagascar. The presence of such reptiles only 250 miles from the African mainland again raises the problem of the continental origins of the island. The key to the enigma may be found in the fact that the iguanas are older than the agamids and preceded the latter in populating many islands. When the agamids appeared Madagascar might already have been separated from Africa, which would explain why in this particular place the agamids never managed to supplant their predecessors or to compete with them.

## Island of harmless reptiles

Similar problems are posed by the presence on Madagascar of a turtle of the genus *Podocnemis,* again a typical inhabitant of South America. Yet there is evidence from fossil remains that representatives of this genus did in fact long ago inhabit parts of Africa, suggesting that the group once enjoyed a much wider distribution than it does today. Four other species of turtle are endemic to Madagascar.

Although there are not many varieties of gecko they are found throughout the island. Most remarkable is the green gecko of Madagascar (*Phelsuma madagascariensis*) which, unlike most of its relatives, is active by day and distinguished by its bright green coloration and rounded pupils.

About half of the world's chameleons are found on the island, ranging in size from tiny reptiles which, even when adult, measure no more than 1¾ inches, to larger ones of up to 24 inches. Many of them have few features in common with African chameleons, having probably evolved in isolation here for millions of years. The Madagascar chameleon (*Chamaeleo oustaleti*) has extremely powerful jaws and a strong tongue which, according to Schmidt and Inger, enable it to catch such prey as rodents and to chew them into small pieces.

There are 23 species of snakes on the island, of which 21 are native; but all of them are either aglyphs or opisthoglyphs, which means that they either possess no venom glands or that if they do their fangs are situated far back in the mouth. Thus they represent no danger either to humans or to large and medium-sized mammals. The fact that there are no cobras,

Most geckos, being nocturnal, have no need for spectacular colours; but the only diurnal member of the family, the green gecko, is an exception. Note the rounded pupils, adapted to bright light.

*Facing page:* All the typical features of an island fauna are present on Madagascar. The chameleons (*above*) and geckos (*below*) are enormously diversified; thus the former include several giant-sized species while the latter can boast one species that is active by day.

A few of the most characteristic species of Vangidae on Madagascar.

vipers or rattlesnakes on Madagascar makes it seem very likely that it was already an island before these more highly evolved snakes acquired their 'murder weapons'.

One great African predator, however, is found on Madagascar, namely the Nile crocodile. Here it lives near the coasts and, since there are no large mammals to hunt, feeds principally on fishes. The crocodile is an excellent swimmer and has often been sighted many miles from African shores. Doubtless what happened was that the odd individual was carried across to the island by powerful ocean currents and that eventually a flourishing colony of the huge reptiles came into being.

Except for crocodiles and other species capable of swimming, it would appear that most of the reptiles of Madagascar have evolved in isolation for thousands of centuries, further proof that the land link, if it existed, must have gone far back in time, probably to the age of the dinosaurs, fossil remains of which have been discovered on the island.

## Birds, past and present

Crossing a stretch of ocean is of course simpler for a bird than for a terrestrial animal. This is why Madagascar is the home of more species of birds than of mammals, amphibians and fishes. Including migrants, there are approximately 220 bird species here, most of them with affinities to African species but some clearly of Oriental origin, which is more surprising. There are four endemic families—Aepyornithidae (elephant birds), Mesitornithidae (mesites), Philepittidae (asities and false sunbirds) and Vangidae (vangas).

The gigantic elephant bird of the genus *Aepyornis*, probably the origin of the legendary roc, is unhappily now extinct. The three species of Mesitornithidae are small ground birds with comparatively poorly developed wings. Related to the cranes, the best-known species is the brown mesite (*Mesoenas unicolor*), an inhabitant of the tropical forests along the east coast. The Philepittidae are also found in the humid eastern woodlands and one of the most handsome members of the family is the wattled false sunbird (*Neodrepanis coruscans*), iridescent blue above and opaque yellow below.

The Vangidae have evolved on Madagascar in the same spectacular fashion as the Darwin's finches of the Galapagos Islands and the honey-creepers of Hawaii. Close relatives of the shrikes, they feed on insects and small vertebrates. Although there are enough similarities to suggest that they are descended from a common ancestor they have developed a number of different shapes and behaviour patterns. Even their choice of prey and hunting techniques vary enormously. The blue vanga (*Leptopterus madagascarinus*), for example, hunts insects among the branches; the helmet-bird (*Euryceros prevostii*) attacks geckos, chameleons, lizards and fledglings; Lafresnaye's vanga (*Xenopirostris xenopirostris*) behaves like a flycatcher; and the tit-shrike (*Calicalicus madagascariensis*) emulates the great tit. As for the sickle-bill (*Falculea palliata*), it scoops out arthropods from cracks in bark with the aid of its curved beak, just like woodpeckers and tree-creepers elsewhere. Finally, the coral-billed nuthatch (*Hypositta corallirostris*) occupies an ecological niche equivalent to that of the nuthatches of the genus *Sitta* in other regions, so much so that they were long thought to be closely related.

Little is known about the reproductive habits of the Vangidae, or indeed of other aspects of behaviour, for they are shy and spend much time in the undergrowth. But they nest in trees and the females lay three or four whitish or greenish eggs, flecked with brown.

Although not a native species, one must mention the long-tailed ground roller (*Uratelornis chimaera*), inhabiting a restricted area, some 125 miles long and 30 miles wide, of sandy plain, almost desert, in the south-west of the island. It is extremely rare and it only needs a brush fire or some other accident or disturbance to bring about its complete disappearance.

Several species of raptors are exclusive to Madagascar. Henst's goshawk (*Accipiter henstii*) is a large bird of prey, much resembling European forms and an inhabitant of humid regions. It feeds on small mammals and birds. The Madagascar banded kestrel (*Falco zoniventris*) is found in woodland as frequently as on open plains, its diet consisting of insects and small reptiles. Other raptors include the Madagascar cuckoo falcon (*Aviceda madagascariensis*), the Madagascar fish eagle (*Haliaeetus vociferoides*) and the Madagascar harrier hawk (*Polyboroides* (*Gymnogenys*) *radiatus*).

The huge elephant bird of the genus *Aepyornis*, which weighed about half a ton, probably vanished from Madagascar before man arrived. Myths of the Middle East mention a bird known as the roc. Possibly the elephant bird originated such tales.

The long-tailed ground roller is nowadays confined to a small piece of arid terrain in south-western Madagascar where it is in grave peril of extinction.

Among the rarest of all the island's birds of prey is the Madagascar serpent eagle (*Eutriorchis astur*) which, despite its name, is not a true eagle. Very shy and secretive, it lives in the northeastern forests and looks rather like a small falcon.

The Madagascar serpent eagle, which is not a true eagle and may not even include snakes in its diet, is one of the rarest and least known birds of prey on the island.

## The fascinating behaviour of lovebirds

The grey-headed or lavender-headed lovebird (*Agapornis cana*), one of a group of birds so named because males and females form lifelong pairs, is a native bird of Madagascar, although its relatives are widely distributed through Africa. Its interest derives chiefly from a series of evolutionary investigations performed under laboratory conditions, suggesting a relationship with the hanging parrakeets of Asia (*Loriculus*) and a descent from a common Oriental ancestor. William C. Dilger, an ornithologist of Cornell University, has established that in certain details of behaviour this and other lovebirds resemble such an ancestor yet have evolved differently, so that the local species and subspecies demonstrate a fascinating variety of behaviour patterns. Based on experiments with all but one of these lovebirds (Swinderen's black-collared lovebird being not available to him) Dilger reconstructed a family tree showing the evolution of the small parrots as represented today on the island. The illustration on the facing page shows that the most primitive species is the

*Facing page:* This diagram of a family tree, based on the work of William Dilger, shows how different species and subspecies of lovebirds, most of them African but one also found on Madagascar, have evolved. Those near the base of the tree are primitive species that stem from Asiatic stock (the hanging parrakeets are descended from the same ancestral forms). Those higher up have evolved more recently. Each group exhibits interesting variations of behaviour.

grey-headed lovebird and that the most recent and highly evolved is the subspecies known as the black-cheeked lovebird (*Agapornis personata nigrigenis*).

The grey-headed lovebird, together with the red-faced lovebird (*Agapornis pullaria*) and the Abyssinian or black-winged lovebird (*Agapornis taranta*) form the main trunk of this family tree from which branch Swinderen's black-collared or Liberian lovebird (*Agapornis swinderniana*), the peach-faced or rosy-faced lovebird (*Agapornis roseicollis*) and four subspecies of the masked or yellow-collared lovebird (*Agapornis personata*). All these birds show a number of behavioural traits indicating a stemming from common stock. Thus the more primitive species live in pairs, like their Asiatic ancestors, whereas the more recently evolved species and subspecies have a pronounced social tendency, often collecting in flocks and nesting in colonies. Such modifications of behaviour at certain seasons also influence the actual formation of pairs. In all species the courtship ceremonial prior to mating is fairly simple and within only a few hours a durable relationship is formed between male and female; but whereas in the more primitive species this does not take place until the birds are about four months old, by which time they are adult and independent, in the gregarious forms it occurs much earlier, when the birds are only two months old and still arrayed in juvenile plumage.

Other aspects of behaviour indicating a progressive change of habit are shown in the way in which the courting birds rub and incline their heads and in intraspecific fighting. In the grey-headed lovebird and two older species the female will, if she wishes to have nothing to do with a male, show her disdain by rubbing her head against his nearer claw. This gesture also occurs among the more highly evolved forms but has taken on a purely ritual significance, divorced from its original meaning. It is then just a part of the courtship display.

There are variations too in aggressive attitudes and intraspecific combats. The more primitive lovebirds, although they seldom confront one another in a hostile manner, fight with surprising violence when they do come to blows; the others confine their rare confrontations to tiny pecks of the beak and nibblings at one another's feet.

In the older species there is distinct sexual dimorphism (as regards colour) which again is not the case with the more modern forms. This lack of differentiation often compels the females of the latter groups to exaggerate their femininity by erecting their feathers to attract the eye of their suitors. The females of the more primitive species do not ruffle their feathers during the breeding season. All females, nevertheless, indicate their disposition for sexual activity by bending forwards and raising the head and tail.

Unlike the majority of Psittaciformes, the lovebirds prepare their nest inside a tree hollow; but here again there is a difference in the manner in which they build. The three species which we have described as primitive, as well as the rosy-faced lovebird, tuck the materials to be used for nest-building among the feathers of their back to transport them, but the remainder

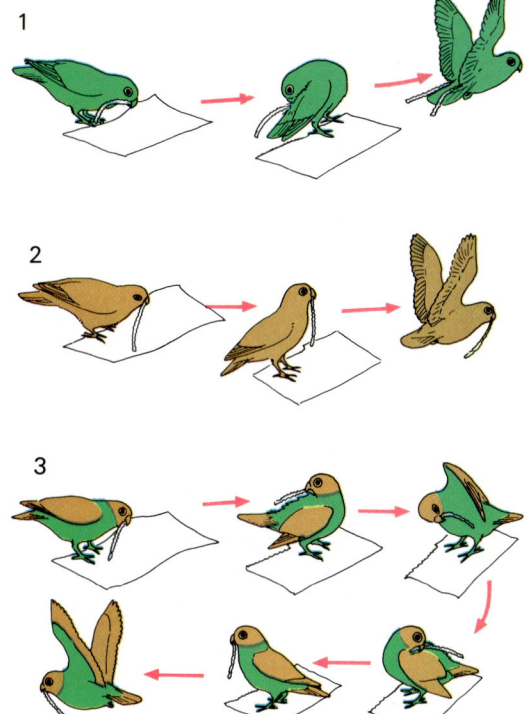

William Dilger's experiments with lovebirds revealed different methods of carrying materials for nest-building. Using pieces of paper, which the birds tore into strips, Dilger found that a rosy-faced lovebird (1) would carry them among the feathers of its back, while a Fischer's lovebird (2) picked the paper up in its beak. A hybrid of these two species (3) showed uncertain behaviour, first attempting to place the strips on its back but eventually flying off with the paper in its bill.

*Facing page:* Lovebirds are of Oriental origin but are now inhabitants of Africa and Madagascar. One subspecies of the former group is Fischer's lovebird.

simply carry bits of grass and twigs in their bill, as do most other birds. Dilger conducted experiments with sheets of paper and was interested to note that the birds all tore them into small shreds to adorn their nests. His purpose was to prove that the manner in which the different species and subspecies carried materials to the nest was hereditary, transmitted from one generation to another. By cross-breeding a rosy-faced lovebird with a Fischer's lovebird (*Agapornis personata fischeri*) Dilger found that the resultant hybrid tried to place strips of paper in the feathers of its back but when these repeatedly fell out the bird, after a number of further futile attempts, eventually picked them up in its bill, revealing a mixture of behaviours characteristic of both parents.

These fascinating experiments with the lovebirds of the genus *Agapornis* provide fresh evidence of the fact that some aspects of animal behaviour, like the more obvious characteristics of anatomy and physiology, may be inherited and that they too can be modified in the course of evolution.

## The multiform tenrecs

Although the lemurs are by far the best known of the mammals of Madagascar, most of the island's mammals are insectivores belonging to the family Tenrecidae. This family–in terms of racial history and development, zoogeography, adaptation and so forth–provides a remarkable lesson in evolution.

There are ten genera of tenrecs on Madagascar, comprising thirty species. Some of these animals look like hedgehogs, others like rats or shrews; but studies in comparative anatomy have shown that, despite structural differences, the tenrecs, more than any other group of living mammals, most closely resemble the primitive forms of insectivores from which the placental mammals later evolved.

The closest relatives of the tenrecs are the Solenodontidae, huge insectivores from the Antilles, and, to a lesser degree, the otter shrew (*Potamogale velox*) of the African rivers and streams. The link between insect-eating mammals from places so far apart as Madagascar and the Antilles seems to pose a zoological problem as tricky to resolve as that of the iguanas. But studies revealing the primitive origins of the tenrecs have led scientists to conclude that at one time these animals were widely distributed in all parts of the world and that only in areas fortunate enough to have severed their land links and gone through a long period of isolation, such as Cuba, Haiti and Madagascar, have such forms survived.

The 'tailless' tenrec (*Tenrec ecaudatus*) looks like a cross between a hedgehog and a porcupine for its skin is covered with hair and sharp defensive spines. Like other tenrecs it is a nocturnal animal, an inhabitant of plains with a scattering of bushes and shrubs rather than of the tropical forest or desert. It burrows with the powerful claws of its forefeet, feeding not only on insects and worms but also (like the hedgehog) on frogs, snakes and lizards. Furthermore, the tailless tenrec is among the most prolific of living mammals; the female, who possesses twenty-

---

### COMMON TENRECS

Class: Mammalia
Order: Insectivora
Family: Tenrecidae
Diet: insects, worms, small vertebrates, varying with species
Gestation: not known, probably short
Number of young: very variable, according to species, normally 4-16

#### 'TAILLESS' TENREC
(*Tenrec ecaudatus*)

Total length: 11-16 inches (27.5-40 cm)
Length of tail: $\frac{1}{2}$-$\frac{3}{4}$ inch (10-16 mm)

Greyish-brown or reddish fur, some individuals darker. Pelage a mixture of true hairs, long and bristly, and spines (which in adults are not very sharp). Young have a row of functional spines along back. Hind legs shorter than forelegs.

#### BLACK-HEADED STREAKED TENREC
(*Hemicentetes nigriceps*)

Total length: $6\frac{1}{2}$-8 inches (16-20 cm), tail vestigial

Pelage black and white striped, giving animals a skunk-like appearance. Row of erectile spines along back and on neck. Underparts whitish. Young have similar stripes and spines and are generally more strikingly marked.

#### LONG-TAILED TENREC
(*Microgale longicaudata*)

Length of head and body: $2\frac{1}{4}$-$2\frac{1}{2}$ inches (6-7 cm)
Length of tail: 6-$6\frac{1}{2}$ inches (15-16 cm)

Shrew-like with disproportionately long tail, naked towards tip. Upper parts reddish-brown or blackish; underparts paler. Protruding ears.

Some typical tenrecs of Madagascar.

four teats, is capable of giving birth in one litter to as many as thirty-one babies. These youngsters learn to run about soon after they are born, opening their eyes after the first week and able to fend for themselves before they are as much as one month old.

The large Madagascar 'hedgehog' (*Setifer setosus*), Fontoynont's 'hedgehog' (*Dasogale fontoynonti*), and the small Madagascar 'hedgehog' (*Echinops telfairi*) all look like real hedgehogs although they cannot roll themselves completely into a ball. Nocturnal animals, they move their spines to and fro to frighten an enemy, as do porcupines. Perhaps because of this they are not much admired by the island's inhabitants who have the notion that an animal which cannot defend itself with its claws must be cowardly. For that reason warriors were at one time forbidden to eat the flesh of these animals.

The banded tenrec (*Hemicentetes semispinosus*), found in most parts of the island, has several close relatives of the same genus. It looks rather like a skunk, with contrasting black and creamy stripes, yet its comparatively long spines are similar to those of a hedgehog. When it is frightened it adopts a defensive posture, moving its spines in a forward direction. These spines are easily detached from its body and stay firmly stuck in that of the enemy. What is more, at night, when a group of these little animals sally forth (each one is about the size of a human hand) the noise made by the spines knocking against one another is enough to keep the members of the group close together. Local people describe how, when one of the animals comes across an earthworm, it dances about and agitates its spines to pass the message to its companions who immediately rush into the area to look for similar tasty morsels.

A second group of tenrecs is made up of individuals which

Fossa

Jaguarundi

The fossa of Madagascar is one of the Viverridae, related to civets and mongooses. Yet it looks more like a member of the cat tribe and there is an interesting example of convergent adaptation in its close resemblance to the jaguarundi of the Americas.

---

**FOSSA**
(*Cryptoprocta ferox*)

Class: Mammalia
Order: Carnivora
Family: Viverridae
Total length: 49-53 inches (125-135 cm)
Length of tail: 26 inches (65 cm)
Diet: birds and small or medium-sized mammals, including lemurs
Number of young: 2-3

Largest carnivore on Madagascar. Cat-like appearance with short, compact, reddish-brown fur. Short snout, large eyes and long, curved, retractile claws. Perianal glands secrete nauseous substance as in other Viverridae. Plantigrade, unlike cats, which are digitigrades.

---

have a long tail and a short, thick covering of fur, without spines. One of the most interesting of these is the very small long-tailed tenrec (*Microgale longicaudata*). Measuring a mere $2\frac{1}{4}$-$2\frac{1}{2}$ inches from tip of snout to base of tail, it is among the tiniest of the world's mammals.

It is surprising that in the same family animals reminding one of hedgehogs should live alongside others that both look and behave like moles but this is the case here. The rice tenrec (*Oryzorictes talpoides*) has minute eyes, soft, compact fur, and strong claws on the forefeet for burrowing.

A striking example of adaptation is to be seen in the web-footed tenrec (*Limnogale mergulus*), whose appearance has changed in response to a modified life pattern. The hind feet are very large and the toes are webbed; and the animal feeds on the fruits, roots and shoots of aquatic plants. It is an extremely rare species, perhaps already extinct.

## Madagascan carnivores

In recent times the common mongoose and one species of wild cat have been introduced into Madagascar but the only native carnivores on the island are six genera of Viverridae which include in their ranks ground and tree species, daytime and nocturnal hunters, omnivores and specialised feeders. But since none of them has been studied in great detail there is little reliable information concerning their habits.

The Madagascan broad-striped mongoose (*Galidictis striata*) is one of the rarest of the group, the female possessing a double uterus. But the most astonishing of the lot is the fossa (*Cryptoprocta ferox*), a curious mixture of mongoose, genet and wild cat, long classified as one of the Felidae. It is unbelievable that an animal as large as a fox, which has been the subject of such a wealth of local story and legend, and which is by no means uncommon, should be such a mystery to zoologists. It is a nocturnal animal and much resembles the jaguarundi or otter cat of North and South America, feeding on mammals and small or medium-sized birds. In the breeding season groups of about six animals are seen together and local people say that they are particularly fierce and dangerous at that time, although the truth is that they are more likely to flee any human approach. The litter consists of two or three babies.

## A vanished world

Madagascar has felt the grievous impact of man's intrusions. Approximately 70 per cent of once-virgin land has been transformed. Nowadays the eastern forest regions have been broken up, older endemic formations having been ousted by vanilla and cocoa plantations. The huge deciduous forests in the western part of the island have given way to an immense treeless steppe. Fires have devastated the sub-desert areas in the south-west, while overpasturing and indiscriminate tree-felling have changed the appearance of the landscape everywhere, with the inevitable tragic consequences to local forms of wildlife.

A number of endemic plant and animal species are in danger of extinction and as the forests are cut back so the lemurs become increasingly rare. Eight of them appear in the IUCN Red Data Book; but others have already vanished.

Until comparatively recently there still existed on Madagascar a giant lemur which stood 5 feet tall, a pygmy hippopotamus similar to the species still found in West Africa, a huge terrestrial tortoise and, of course, the enormous flightless elephant bird which, judging by fossils of bones and eggs, still flourished less than a thousand years ago and must therefore have lived alongside man.

## The Seychelles and Mascarene Islands

Although not part of the Malagasy Republic, the Seychelles and Mascarene Islands of the Indian Ocean are close enough geographically to Madagascar for their flora and fauna to exhibit similar links both with Africa and the Orient. Yet both island groups are sufficiently different in structure to be considered separately. If the hypothetical continent of Lemuria really existed it is feasible that the many tiny islets of the Seychelles as well as the fewer but larger islands of the Mascarene group could be the vestiges of this sunken land mass. But, as we have said, the theory is based on pure supposition and there is no scientific evidence that an immense land of lemur-like animals, of the Tertiary epoch, was ever a reality.

Study of the Seychelles archipelago, consisting of 92 islands or islets lying about 600 miles north-east of Madagascar, has

The tenrecs are the principal insectivorous mammals on Madagascar. Some of them, such as the small Madagascar 'hedgehog', are covered with spines and look much like real hedgehogs.

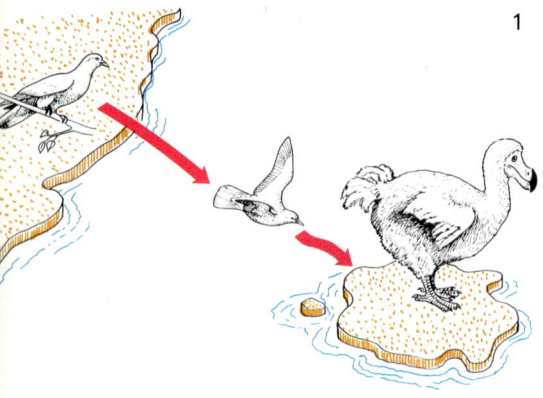
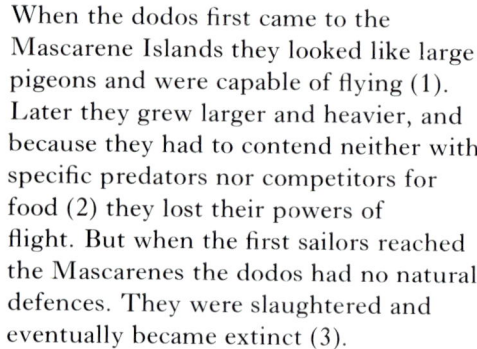
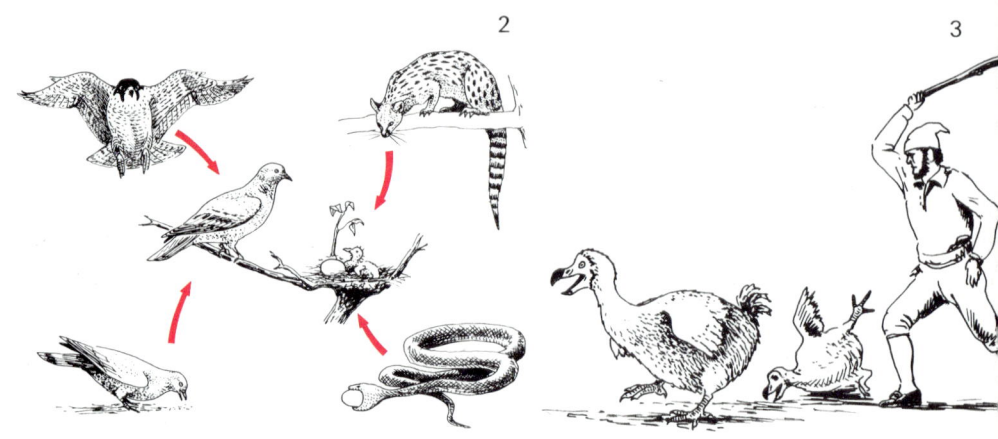

When the dodos first came to the Mascarene Islands they looked like large pigeons and were capable of flying (1). Later they grew larger and heavier, and because they had to contend neither with specific predators nor competitors for food (2) they lost their powers of flight. But when the first sailors reached the Mascarenes the dodos had no natural defences. They were slaughtered and eventually became extinct (3).

revealed the presence of rocks such as granite, syenite, schist, etc, which are known as 'continental' because they are found both on continents and on the islands assumed once to have been joined to them. But this in itself is not enough to prove the existence of Lemuria, because in order for such continental-type rocks to be found on nearby islands all that would have been needed was a narrow bridge of land which at one time was not completely submerged. Nevertheless, whether or not the Seychelles were originally linked with the African mainland, it is obvious that, as on Madagascar, both the plants and animals of these islands exhibit certain typical characteristics of ancient derivation, consistent with a state of isolation, surrounded only by ocean, for a considerable period.

One of the most remarkable features of these islands is the total absence of freshwater fishes and of mammals, apart from flying foxes, bats of the genus *Pteropus*, clearly of Oriental origin, for which vast expanses of ocean would have posed no obstacle. The fauna of the Seychelles consists of amphibians (spadefoot toads of the family Pelobatidae, probably of African origin), a few lizards, snakes and chameleons—some from Africa, others from Asia—and seventeen species of birds, of which fourteen are endemic. Formerly there were giant tortoises of the genus *Geochelone*, which are today extinct, and two birds, a species of white-eye (*Zosterops semiflava*) and the Seychelles parrot (*Psittacula eupatria wardi*) are also extinct.

The Mascarene Islands comprise Réunion, Mauritius and Rodriguez and lie 425 miles, 600 miles and 865 miles respectively east of Madagascar. These are definitely oceanic islands of volcanic origin, although on Mauritius there are schist formations similar to those on the Seychelles, suggesting that in the distant past the island must have been much larger and higher. In these islands too there are neither freshwater fishes nor mammals (again apart from bats); nor are there any amphibians. The only reptiles, both found on Mauritius, are two snakes, both boas, (*Casarea dussumieri* and *Bolyeria multicarinata*) and the green gecko of Madagascar (*Phelsuma Madagascariensis*), which sometimes measure over 10 inches, possibly further proof of the type of gigantism which is so often found on islands.

As for birds, if one omits doubtful types, these consist of twenty-eight species, but many of them are now extinct as the result of human colonisation. The list of birds includes gulls, rails and, most interesting of all, a complete family of birds known as Raphidae, the last representatives of which disappeared soon after man reached these islands.

## Dead as the dodo

Many of us recognise the dodo as one of the animals in *Alice in Wonderland,* and the phrase 'dead as the dodo' enters frequently into our daily conversation. But this bird was no mere figment of Lewis Carroll's imagination (vividly portrayed in Sir John Tenniel's illustrations). Such a bird really existed as recently as three hundred years ago.

In fact the dodo (*Raphus cucullatus*) was only one of three related species of the family Raphidae which all appear on the sad list of extinct animals. The dodo proper lived on Mauritius while its relatives, the Rodriguez solitaire (*Pezophaps solitaria*) and the Réunion solitaire (*Raphus solitarius*) inhabited the other two islands of the Mascarene group. All three species were the victims of hunting, principally serving as food for the sailors of passing ships and being described contemptuously as ugly, lazy and apparently stupid, because when hunted they neither fled nor put up a fight but apathetically allowed themselves to be slaughtered. Their extinction was hastened by the marauding activities of pigs and monkeys introduced by the Portuguese, these animals destroying both eggs and chicks. The dodo vanished from Mauritius and the world in 1681. The Réunion solitaire followed in 1746 and the Rodriguez solitaire survived until 1791. All we know of these species is derived from sketches by naturalists of the 17th and 18th centuries and a few skeletons in museums which enabled zoologists to reconstruct the birds as they must have been in their heyday.

This extraordinary family was eventually classified as one of the Columbiformes (pigeons and doves) although some authors such as Lüttschwager prefer to include them among the rails (Rallidae). All three species evidently looked something like pigeons although they were much larger and sturdier, nearer the size of turkeys. The huge head was duck-like but with an enormous hooked beak. Wings were vestigial so that the birds could not fly. The tail, as is often the case with ground birds, was exceptionally short and in fact little more than an ornamental tuft, without any real function. The feet were stronger than is customary for birds of such dimensions. The reason that these species were not adapted to flying is probably that the absence of ground predators gave total security, which was indeed the case before man put in an appearance.

The dodo evidently had greyish-brown plumage, as did the Réunion solitaire. The Rodriguez solitaire, however, exhibited sexual dimorphism, the male being chestnut, the female completely white, with yellower beak and feet.

It is possible that live specimens were brought back to Europe at the beginning of the 17th century but it seems that none of them survived in those more temperate latitudes. The only stuffed specimen of a dodo, in Oxford, was burnt because of its poor state of preservation.

Dodo (*Raphus cucullatus*)

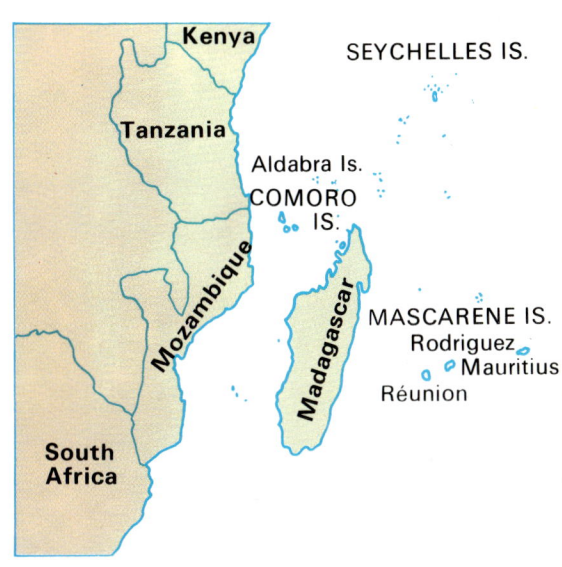

Madagascar and neighbouring islands.

# CHAPTER 105

# Land of the lemur, indri and aye-aye

The lemurs of Madagascar are of particular interest to science for although they are Primates they belong to the suborder Prosimii, indicating their primitive origins. In fact they precede monkeys on the ladder of evolution. This does not mean that they are the ancestors of modern monkeys any more than it is true that man is directly descended from apes. Lemur, monkey and man alike stem from a common ancestor, a primitive form of insectivore; but the many primitive descendants of this mammal became diversified, each being characterised by an ascending degree of evolution.

What then is the structural level of the Lemuriformes (lemurs and their close relatives)? Obviously their development was arrested somewhere along the line. In their case the sense of smell is more important than that of vision, and the muzzle is elongated (giving them a fox-like appearance). The thumb on either hand and the big toe of each foot are opposable to the other four digits. The second toe of each hind foot is furnished with a long claw that is often used for cleaning the fur; and the remaining toes and fingers have flat nails. The cerebral cortex is poorly developed, indicating a low level of intelligence. Reproduction is seasonal and in contrast to the higher Primates the females are sexually receptive only for a short period at one particular time of the year. But among features regarded as indicative of more advanced development are the rudiments of temporal lobes, the trace of a yellow spot on the retina and the elaborate social behaviour of certain species.

Isolated on the island of Madagascar for perhaps fifty million years, the modern lemurs are probably little different; from that original ancestor from which they and other Primates sprang. It is therefore interesting to study the social and family

*Facing page:* One of the most spectacular and well known Lemuriformes is the ring-tailed lemur.

behaviour of a group of animals which, in many ways, must be similar to that distant ancestor of man himself.

Directly related to the various species of loris, the bushbabies and the tarsiers, the Lemuriformes are usually classified in three families–Lemuridae, Indridae and Daubentoniidae. All of them live exclusively on Madagascar, apart from one species inhabiting the Comoro Islands.

## Home of the lemurs

The Lemuridae are the true lemurs, found only on Madagascar. According to the classification adopted by Jean Jacques Petter, the French naturalist who has studied most of these animals in the wild and to whom we owe most of our information, they are divided into eleven genera and twenty species.

Having diversified in the course of evolution, the lemurs of Madagascar occupy the same ecological niches as the monkeys of Africa, Asia and America. Some are the size of mice while others are more than 3 feet long. They include diurnal and nocturnal forms, solitary and social species, frugivores, insectivores and omnivores. Some walk on two feet, some on four. There are both ground and tree species and they have adapted to such contrasting habitats as the tropical rain forests of the interior and the thorny woodland of the west coast. It is true to say that there is hardly any life-sustaining biome that has not been explored and exploited by the lemurs, their former wide distribution having been facilitated by the absence of competition from more highly evolved mammals.

The lesser mouse lemur (*left*) and the dwarf lemur (*right*) are solitary animals which spend the day in a hole or tree hollow and emerge at night to look for the insects and fruit which constitute their food.

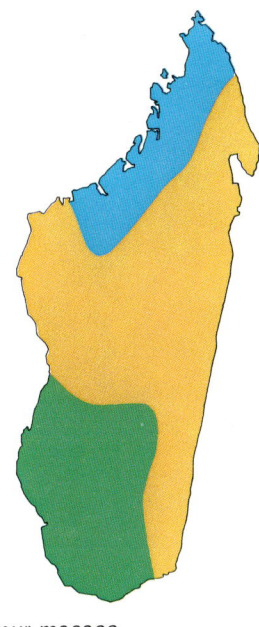

 *Lemur macaco*
 *Lemur catta*

Geographical distribution of the black lemur (*Lemur macaco*) and ring-tailed lemur (*Lemur catta*).

The lesser mouse lemur is seldom seen by day. It seems likely that although it generally shelters in natural cavities this tiny animal, the smallest of all Primates, builds nests in the branches.

---

**COMMON LEMURS OF MADAGASCAR**

Class: Mammalia
Order: Primates
Family: Lemuridae
Diet: Leaves and fruit
Gestation: 4½ months
Number of young: usually one

**RING-TAILED LEMUR**
(*Lemur catta*)

Total length: up to 40 inches (100 cm)
Length of tail: 20 inches (50 cm)

Back grey; abdomen and inside of legs somewhat paler. Pointed muzzle, black at tip; white mask on face and ears; dark surrounds to eyes. Tail white with fifteen brown rings.

**BLACK LEMUR**
(*Lemur macaco*)

Total length: 30-38 inches (75-95 cm)
Length of tail: 16 inches (40 cm)

Sexual dimorphism. Male very dark, almost black; female reddish-brown, darker on hands, feet and face. Forehead and ears very hairy, black in male, white in female. Young have covering of thin, dark grey fur at birth.

In the morning both ring-tailed lemurs and sifakas sit on a branch exposing their bodies to the warmth of the sun.

Dwarf lemurs alternate between periods of frantic activity and phases of complete lethargy. They can go without eating for some time, surviving on the fat stored at the base of the tail.

The lesser mouse lemur (*Microcebus murinus*) is to be found almost everywhere and is not only the smallest member of the family but also the smallest of all Primates. It belongs to the subfamily Cheirogaleinae, which also includes the dwarf lemurs of the genus *Cheirogaleus*. Two species of the latter live in the eastern forests and a third in the south-western parts of the island. Another member of this subfamily is the fork-marked mouse lemur (*Phaner furcifer*) which is 2 feet in length but whose distribution and behaviour is little known.

The subfamily Lemurinae comprises the true lemurs, some of which display a fairly elaborate form of social behaviour. This is certainly true of the genus *Lemur* and to a lesser extent of the genera *Varecia* and *Hapalemur*. The members of the genus *Lepilemur*, however, are staunch individualists.

There are no distinct subfamilies among the Indridae, but the family consists of three genera—*Propithecus* (sifakas or monkey lemurs), *Avahi* (woolly lemur) and *Indri* (indri or dog-faced lemur). These animals usually stand upright and move in a series of long jumps from branch to branch. The sifakas organise themselves in family groups. The other two species appear to have gregarious habits too but the social structure of their groups is slightly different.

The third family, Daubentoniidae, is made up of one species only, the aye-aye (*Daubentonia madagascariensis*). It is a highly specialised animal and some authors classify it in a separate group distinct from the lemurs.

## Small nocturnal lemurs

The lesser mouse lemur, the dwarf lemurs and the fork-marked mouse lemurs are nocturnal animals, usually spending their day concealed in holes and tree hollows, becoming active only at night. Yet these species apparently build their own refuges. Petter once saw two dwarf lemurs in a nest of leaves, hidden in the foliage more than 10 feet from the ground. It appears that the lesser mouse lemur also builds nests of this kind.

The lesser mouse lemur moves along the branches like a tiny rodent, with a series of rapid little steps interspersed with sudden halts and equally unexpected short leaps, similar to those of a dormouse. It is an omnivore, feeding principally on fruit and insects; and, as is the case with many Asiatic prosimians, it probably includes eggs and nestlings in its diet as well.

The dwarf lemurs also move like mice but they are somewhat heavier and rather clumsier than the lesser mouse lemur. Their food too consists in the main of insects and fruit. These animals frequently feed while squatting on their hindquarters so that the hands can be used for conveying food to the mouth.

In particularly dry weather the dwarf lemurs fall into a kind of lethargic state, not unlike hibernation. Their body temperature drops and they do not move around for food, surviving on the fat stored at the base of the tail. Petter notes that this phenomenon never occurs when the outside temperature is high, and such a state of torpor, according to Kolar, is unrecorded among the lemurs of the Vienna Zoological Gardens.

Aye-aye
(*Daubentonia madagascariensis*)

Black lemur (male)
(*Lemur macaco*)

Indri
(*Indri indri*)

Diademed sifaka
(*Propithecus diadema*)

Black lemur (female)
(*Lemur macaco*)

As a general rule, these lemurs of the subfamily Cheirogaleinae give birth to two or at most three babies, after a gestation of 60-70 days. The mother bites through the umbilical cord and swallows the placenta. The babies, already covered with hair, weigh only 3-4 grammes. The dwarf lemurs open their eyes after one day, the lesser mouse lemurs after four days.

Apparently the mother never carries her young on her back or clasped to her belly, but carries them from place to place in her mouth for about three weeks, by which time they are agile runners and jumpers. At two months of age baby lesser mouse lemurs already look adult but it is a year before they are capable of reproducing.

During the night, when they are active, it is rare to see more than two or three of these small animals together, for they are solitary by habit. In the daytime, however, several of them may shelter in the same hole. This suggests that they are not true individualists in the sense that each defends its own territory. Unlike other lemurs, which defend their boundaries by emitting loud cries, rather like the gibbons of Asia and the howling monkeys of South America, these little lemurs do not seem to care about territory at all.

The sense of smell plays an important role in their life. As they move along they generally keep their snout close to the branch, and it seems that they find their way to fruit as much by scent as by vision.

The black lemur, like the ring-tailed lemur, marks the branches demarcating its territory with strong-smelling secretions from glands in the wrists and arms as well as from perianal glands.

Verreaux's sifaka
(Propithecus verreauxi)

Ring-tailed lemur
(Lemur catta)

Woolly indri
(Avahi laniger)

Although some lemurs leave their young in the nest others carry them around until they are independent. When it is very small it clings to its mother's belly; later she may carry it on her back.

*Facing page:* The ring-tailed lemur (*above*) and the black lemur (*below*) both exhibit a fairly elaborate form of social behaviour. Although they have few natural enemies apart from man, both species are well able to fend for themselves, the former by emitting an alarm cry which is immediately taken up by its companions. In the latter species there is marked sexual dimorphism. The female, shown in this picture, is reddish-brown whereas the male is almost black.

# The sociable ring-tailed lemur

The cat or ring-tailed lemur (*Lemur catta*) is one of the more social species, going about in groups of up to two dozen individuals. At an ecological level they may be compared to baboons for they spend some, but not all, of their time on the ground. Basically, like all lemurs, they are arboreal but according to the observations of Alison Jolly, the young American naturalist who has made a detailed study of the species, they spend only 35 per cent of the time off the ground, at a height of 40-70 feet.

The ring-tailed lemur is markedly territorial. One area of about 25 acres, studied by Alison Jolly, contained only a single colony comprising twenty animals in 1963 and twenty-five in 1964. They slept together in trees but seldom remained for two consecutive nights in the same spot. As soon as the sun rose they moved off in search of food.

The animals appear to eat only fruit. According to Hill, the diet consists of bananas, wild figs and, above all, the fruit of the prickly pear, a cactus of the genus *Opuntia,* the tough skin of which they tear off with their teeth. After eating, and sometimes even during the meal, they take their daily sun-bathe. Sitting on the higher branches, they open arms and legs wide, exposing as much of the body as possible to the sun's rays, then turn round and warm their back.

Before retiring to sleep they emit noisy cries which carry some distance and probably serve to inform neighbouring troops of the location of their dormitory so that no conflict arises.

Ring-tailed lemurs are rather more active than other species despite the fact that for long periods during the day they do not budge at all. A single journey may take them a couple of hundred yards and they will cover three times this distance in the course of the day. As a rule the young females and the strongest males lead the way, followed by the adult females and young. In almost all the groups studied there were more males than females. Naturalists have noted the same proportions among captive lemurs and it would appear that this is necessary for social life and reproductive behaviour to proceed normally.

## BEHAVIOUR OF CERTAIN MALAGASY LEMURS
(according to Jolly, Petter and Rousseaux)

|  | Time of activity | Diet | Size of group | Rearing of young |
|---|---|---|---|---|
| **FAMILY LEMURIDAE** | | | | |
| **Subfamily Cheirogaleinae** | | | | |
| Lesser mouse lemur (*Microcebus murinus*) | Night | Insects and other small animals; fruit | 1-8 | In nest |
| Dwarf lemurs (genus *Cheirogaleus*) | Night | Insects and other small animals; fruit | 1 | In nest |
| Fork-marked mouse lemur (*Phaner furcifer*) | Night | Insects and other small animals; fruit | 1-3 | ? |
| **Subfamily Lemurinae** | | | | |
| Weasel-lemur (*Lepilemur mustelinus*) | Night | Leaves and fruit | 1-2 | Usually in nest |
| Grey gentle lemur (*Hapalemur griseus*) | Twilight | Leaves and fruit | 6 | Carried by mother |
| Ring-tailed lemur (*Lemur catta*) | Twilight | Leaves and fruit | 12-24 | Carried by mother |
| Black lemur (*Lemur macaco*) | Twilight | Leaves and fruit | 6-15 | Carried by mother |
| Mongoose lemur (*Lemur mongoz*) | Twilight | Leaves and fruit | 6-8 | Carried by mother |
| Red-bellied lemur (*Lemur rubriventer*) | Twilight | Leaves and fruit | 4-5 | ? |
| Ruffed lemur (*Varecia variegata*) | Twilight | Leaves and fruit | 2-4 | In nest |
| **FAMILY INDRIDAE** | | | | |
| Woolly lemur (*Avahi laniger*) | Night | Leaves, fruit and bark | 2-3 | Carried by mother |
| Verreaux's sifaka (*Propithecus verreauxi*) | Day | Leaves, fruit and bark | 2-10 | Carried by mother |
| Indri (*Indri indri*) | Day | Leaves, fruit and bark | 3-4 | Carried by mother |
| **FAMILY DAUBENTONIIDAE** | | | | |
| Aye-aye (*Daubentonia madagascariensis*) | Night | Insects and other small animals; fruit | 1 | In nest |

Over the bare patches of skin on wrists and inner arms there are glandular protuberances which in the males are present also in the armpits; and both sexes mark the places which they regard as their particular haunts with the secretions of the perianal glands. The lemurs frequently sit with their long tail cradled in their arms, rubbing it against their wrists so as to cover it with these strong-smelling secretions. In this way the tail can serve both as a visual and an olfactory signal.

May is the breeding season when the animals form pairs, male and female sleeping huddled against each other in the foliage. Since a number of males may be similarly inclined, it is probable that a female will mate with more than one partner. Just the same, the original pairs seem to remain together throughout the mating season, although it has not been possible to determine whether the identical individuals come together season after season. Male and female lick and sniff each other

and undertake mutual grooming during this period.

After a gestation of four and a half months the female gives birth to one baby, and occasionally twins. Hill cites the case of a mother who produced two babies and promptly ate one, rearing the other normally. When the baby is born it is reasonably well developed, firmly grasping the hair of the mother's abdomen and being carried by her lengthwise. It feeds alone when it is one month old but continues suckling for four or five months. It is independent at six months and mature at a year.

## Territory of the black lemur

The social behaviour of the black lemur (*Lemur macaco*), which comprises a number of distinct subspecies, has been studied by

The ruffed lemur is the least known of the Lemuridae for it is an inhabitant of dense forests and emerges only in darkness. It feeds on leaves and fruit and lives in family groups.

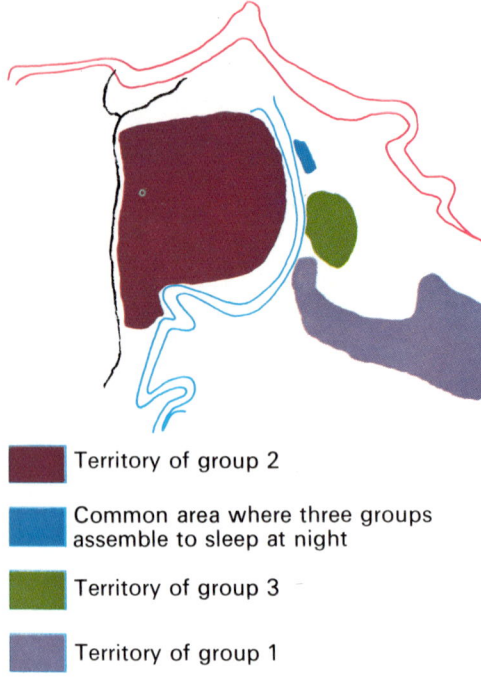

■ Territory of group 2
■ Common area where three groups assemble to sleep at night
■ Territory of group 3
■ Territory of group 1

Territories occupied by three groups of black lemurs, studied by J. J. Petter, in the northern part of the island of Nosy-Komba.

Petter on the islet of Nosy-Komba and in the Lokobe reserve. Groups consist of six to fifteen individuals, including young. There are almost always more males than females but the latter have a high social standing and it is invariably a young female who leads the way when the group is on the move.

Each troop lives in a well defined area which it defends vigorously and which can be regarded as its territory in the accepted sense. Yet, surprising as it may seem, Petter has seen two or three neighbouring troops leaving their daytime territories to assemble together in a common sleeping place. Here they will mingle indiscriminately, so that it is virtually impossible to sort them out. On Nosy-Komba Petter came across four dormitories of this type, three of them situated on the seashore, accommodating ten different troops of black lemurs.

The colonies thus formed are extremely noisy. Fights are continually breaking out and there are mad pursuits through the branches, all to the accompaniment of strident cries. This species is notable for the frequency and variety of its calls. One especially characteristic sound is a kind of loud snarling, repeated almost continuously when a group is on the move or feeding, and rising in pitch as a warning signal when it is emitted simultaneously by all the members of the troop. Another special sound is given out when the lemurs settle down on their individual branches to sleep.

Lemurs have few natural enemies, apart from man. Even the fossa only attacks very young lemurs. Eagles and sparrowhawks sometimes launch an attack but against these raptors the lemurs have evolved an effective form of defence. When a bird of prey ventures uncomfortably close, the first lemur to spot it lets out a sequence of alarm cries which are immediately taken up by all the members of the troop and even, more remarkably, by other animals in the vicinity. The result is an unbelievable hubbub of sound which usually frightens the enemy off. The behaviour of ring-tailed lemurs under such circumstances puts one in mind of various species of small Holarctic birds which, on sighting a bird of prey, mob it and let out warning cries, thus alerting not only their companions but also other animals in the area which may be potential victims.

## The sifakas

Verreaux's sifaka (*Propithecus verreauxi*) is an inhabitant of the tropical forests of western Madagascar and the diademed sifaka (*Propithecus diadema*) of the eastern forests. Both species live in groups of four or five individuals, as a rule.

Like the lemurs of the genera *Indri* and *Lepilemur* these so-called monkey lemurs tend to assume an upright position both when moving about on the ground and when at rest. Their hind feet are much longer and more powerful than their arms. As a rule they get around by making long leaps from one tree to another, the body being held horizontally. If pursued a sifaka can jump more than 30 feet, but whereas the weasel-lemur (*Lepilemur mustelinus*) uses its tail to keep its balance when jumping, the sifakas simply hold it loosely for it is not very

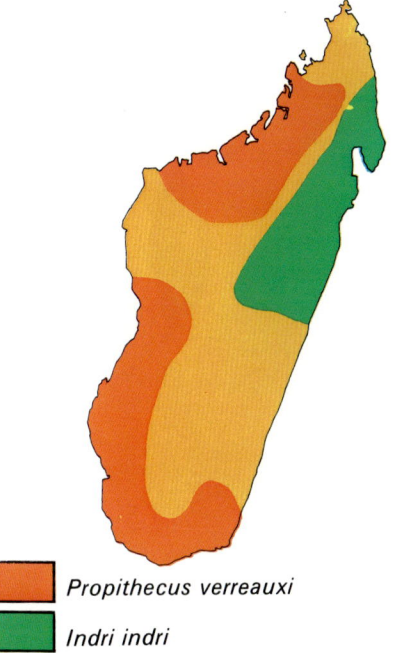

Geographical distribution of Verreaux's sifaka (*Propithecus verreauxi*) and the indri (*Indri indri*).

---

**COMMON MALAGASY INDRIDAE**

Class: Mammalia
Order: Primates
Family: Indridae
Diet: leaves, fruit and bark
Gestation: 5 months for sifaka
Number of young: usually one

**VERREAUX'S SIFAKA**
(*Propithecus verreauxi*)

Total length: 40 inches (100 cm)
Length of tail: 21½ inches (53 cm)

Fur variable in colour. Naked face; well developed eyes. Forelegs longer than hind legs. Long, poorly muscled tail.

**INDRI**
(*Indri indri*)

Total length: up to 40 inches (100 cm)
Length of tail: 2-2½ inches (5-6 cm)

Largest of living Lemuriformes. Fur variable in colour with mixtures of grey, black and white. Head, arms, back and chest black. Back of legs, rump and tail white. Face black. Eyes yellow.

---

The female sifaka is one mother who does not leave her baby alone while it is very young but carries it around with her wherever she goes.

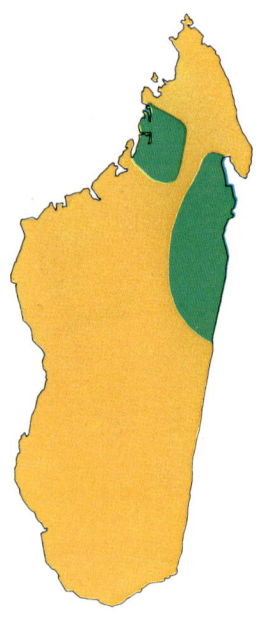

Geographical distribution of the aye-aye.

**AYE-AYE**
(Daubentonia madagascariensis)

Class: Mammalia
Order: Primates
Family: Daubentoniidae
Total length: 34-40 inches (86-100 cm)
Length of tail: 22 inches (55 cm)
Weight: almost 4½ lb (2 kg)
Diet: mainly insects and fruit
Gestation: unknown
Number of young: one

Fur usually dark brown or black; facial mask and neck sometimes yellow. Bushy tail. Hands and feet black. Toes and fingers very long and slender, especially the middle finger which is furnished with a curved claw. Only the thumb has a flat nail. Large eyes and ears.

muscular. The tail, incidentally, plays an even more insignificant role in the indri (*Indri indri*) which, although the largest of all lemurs–over 3 feet–has a mere stump for a tail.

Alison Jolly reports that sifakas move about very little during the day, but Petter points out that journeys may vary in length and frequency according to the outside temperature.

Leaving its dormitory, the troop may travel several hundred yards to a convenient site for sun-bathing and as far again in quest of food. After the meal there is a long siesta lasting until late afternoon; then the sifakas take another evening meal before retiring to sleep. When not feeding they engage in meticulous grooming sessions. This is characteristic of all lemur species, the fur being cleaned with the aid of the lower incisors and canines which are approximately of equal length and which project forwards, forming a kind of comb.

Although territories sometimes overlap, each troop, in Alison Jolly's opinion, retains a central area which is jealously guarded against members of other groups. In dense forest the size of a piece of territory will be restricted, but where vegetational growth is sparser the domain will be considerably more extensive. All depends on food availability.

Although apparently slow and not very nimble, except when climbing vertical trunks, sifakas sometimes hang upside-down to reach fruit and may even swing along a branch by their arms. The entirely vegetable diet consists mainly of leaves, fruit and tender shoots. Normally they grab food directly with the mouth but they can use their hands when seated. It is very rare for them to be seen drinking, except when licking droplets of water from leaves after a fall of rain.

The mating season is from January to March and gestation lasts five months. Petter states that only one baby is born every year in a sifaka family group but Jolly, although agreeing that one or two is the norm, reports having seen three juveniles in one group. Like other lemurs the female sifaka carries her baby around with her, clutched lengthwise across her abdomen. It is not known for certain whether within a single family group adult males engage simultaneously in sexual activity.

## Nosy-Bé, last refuge of the aye-aye

It is interesting to note that in Ancient Rome the spirits of the dead were called *Lemures* and indeed these largely nocturnal animals with their noisy screeching and enormous eyes have something ghost-like about them. The strangest member of the family is the aye-aye (*Daubentonia madagascariensis*), worshipped by local people for centuries and today in danger of becoming extinct. It is in fact the only surviving representative of a once widely distributed family and may be the oldest of living Primates, as primitive as the tree shrews (*Tupaia*).

The aye-aye looks like a large squirrel and was for some time classified–because of its general appearance and the shape of its incisors–as a rodent. But the most remarkable features of its anatomy are the limbs, especially the hands. Only the thumb has a flat nail; the other toes and fingers are exceptionally long

and slender, with claws, and the middle finger of each hand, longest and thinnest of all, ends in a curved claw which is used, like the beak of a woodpecker, for scooping grubs from bark and for other feeding purposes.

In searching for food the aye-aye, it has been said, taps its middle finger against the trunk or rotting branch in order to locate the larvae of xylophagous insects, then scoops them out. It also uses this finger to extract the pulp of fruits; and when it drinks it bends the head low, until almost touching the surface, then quickly flicks the water with its finger into the mouth.

The aye-aye's diet is very varied, for in addition to vegetable substances it feeds on animal matter. Insect larvae are favourite items but fruit, eggs, honey and even fledglings are also eaten. Exclusively nocturnal, it builds a nest in which it sleeps during the day and where it eventually raises its young.

The tragic situation of the aye-aye arises from the fact that its habitat has been almost totally destroyed. The eastern rain forests where it used to live, from sea level up to 2,000 feet and above, are almost gone. The World Wildlife Fund and the International Union for the Conservation of Nature and Natural Resources, on the initiative of J. J. Petter and M. Vadon, have therefore created a reserve for the surviving aye-ayes on the small island of Nosy-Bé, off the north-west coast of Madagascar, where suitable conditions are still to be found. Four males and five females were let loose in the reserve and it is hoped that their descendants will prove sufficiently fertile to ensure the survival of this fascinating species.

Nosy-Bé is the last refuge of the aye-aye but other lemurs are also imperilled for similar reasons. The indri, the sifakas, the fat-tailed mouse lemur, the fork-marked mouse lemur, the mongoose lemur and the western subspecies of woolly lemur are all listed in the IUCN Red Data Book of threatened mammal species.

The hand of the aye-aye is a remarkable tool. The long toes of the hind foot (*right*) all have pointed claws, but the hand is structured quite differently. The thumb is furnished with a flat nail and the middle finger, longer than the others, has a particularly sharp, curved claw. This middle finger can be put to a variety of uses, such as scooping grubs from bark, extracting pulp from fruit and flicking drops of water to the mouth.

CHAPTER 106

# The evolutionary curiosities of the Galapagos Islands

About 600 miles west of the coast of South America, almost on the equator, the thirteen large islands and numerous smaller islets that make up the archipelago known as the Galapagos are washed by the waters of the world's largest ocean, the Pacific.

The total area of these islands, now part of Ecuador, is a little over 3,000 square miles. Half of this area is occupied by the largest island, Isabela, which is also the highest, with peaks of up to 5,600 feet. The Galapagos Islands are all of volcanic origin and many parts are stark lava deserts. They were never linked to any continent and simply emerged from the sea bed as a result of immense submarine upheavals.

In the course of thousands of years winds and ocean currents carried certain plants and animals to these deserted shores. Many of them found conditions suitable for their survival and proliferated, forming new branches that evolved in quite a different manner from that of their ancestors because of the isolated conditions in which they developed.

For centuries no man disturbed the peace of these lost islands. But on a February day in 1535, a ship carrying Bishop Tomas de Berlanga sailed from Panama for Peru. For eight days wind and waves buffeted the ship. Then the storm abated, leaving the vessel drifting helplessly southwards through calm waters, well out of sight of the South American mainland. Food and water were almost exhausted when on March 10 the distant shape of mountains loomed through the mist and fog. Soon the ship was anchored off the shores of a stony, cactus-dotted wasteland. But the desperate sailors found fresh water to drink and having recovered their strength began to pay closer attention to the remarkable animals of the lava-strewn desert–huge lizards diving lazily into the sea, giant tortoises ambling ponderously

*Facing page:* Few birds are capable of using tools for feeding. The Egyptian vulture smashes eggs open with rocks or stones, and the woodpecker finch, shown here, an inhabitant of the Galapagos Islands, uses a cactus spine to dislodge insects from cracks in bark.

The Galapagos Islands, 600 miles west of the coast of South America, have never been linked to any continent and are of volcanic origin. Winds and ocean currents have carried plants and animals to the shores of the various islands and these have developed in isolation to assume many strange and wonderful forms.

Most of the birds of the Galapagos Islands, unthreatened by terrestrial predators, have lost their powers of flight. One such species is the flightless cormorant.

across the beaches, sea elephants basking unconcernedly in the sun. There were also penguins, birds of prey that allowed themselves to be touched, and other strange animals that showed no fear of humans. So impressed were the Spaniards that they called the islands Las Encantadas—The Bewitched. Later they were named Galapagos, after the Spanish word for tortoise.

## Pirates, whalers and scientists

After their accidental discovery the Galapagos Islands knew no more peace. The islands were transformed into hideouts for South Sea pirates, trading and provisioning posts for shipping, bases for whalers. No person who visited these lonely islands could fail to be astonished at the surprising range of wildlife that still flourished there; but although British seamen such as Lionel Wafer, William Dampier and Woodes Rogers provided colourful and fairly accurate descriptions of the scenery and the strange plants and animals of the Galapagos Islands, it was many years before the unique flora and fauna became the focal points of scientific interest.

Exactly three hundred years after the visit of Bishop de Berlanga and his crew, another ship put in at the Galapagos Islands. She was the 242-ton vessel H.M.S. *Beagle* and the date was September 15, 1835. The *Beagle* had sailed from Plymouth on December 27, 1831. She had touched at various islands in the Atlantic and at places along the coasts of Brazil, Argentina, Chile and Peru; and after calling at the Galapagos Islands she was to spend another year crossing the South Pacific and visiting New Zealand, Australia and the islands of the Indian Ocean before returning home, via Africa and South America again, after a round-the-world voyage lasting five years. On board the *Beagle* was a naturalist, Charles Darwin.

Darwin's visit to the Galapagos Islands proved to be an epoch-making event. For it was largely as a result of his studies of the local wildlife that he was able to develop his theory of evolution and natural selection, later to be published in his famous book *The Origin of Species*. Darwin's theory is now universally accepted and has been reinforced by subsequent scientific investigation.

## Pacific crossroads

The unique importance of the flora and fauna of the Galapagos Islands derives from the fact that the archipelago is situated at the centre of a marine 'crossroads' formed by the convergence of two contrasted systems of ocean currents. Here the warm, clear waters of the equatorial counter-current flow in from the west, clashing with the cold Humboldt current from the east. The latter bathes the shores of Peru and Chile and then swings north-westward across the Pacific. It was this current which swept Bishop de Berlanga's ship towards the Enchanted Isles in the 16th century. The confluence of warm and cold waters is responsible for the exceptional wealth of marine life, which in turn explains the abundance of birds and sea mammals on and around these desolate islands.

The cold Humboldt current has a profound effect on the climate so that the west coast of Peru and the shorelines of the Galapagos Islands are characterised by the same type of extremely dry conditions. On all the islands of the group, progress in the coastal belt is difficult, not only because of the almost intolerable drought and heat but also because of the immense obstructing blocks of lava that are scattered everywhere and the thick clumps of spiny cactus which in some places are so tall as to form impenetrable barriers.

The really determined explorer who decides to brave the heat and sets off towards the top of a nearby mountain will see the clear dividing line between the dry, narrow coastal strip and the first transitional zone of vegetation, rising from 100 feet to about 650 feet, consisting of prickly pear, interspersed with occasional trees. As he climbs he will find that the cacti gradually yield place to trees of the genus *Scalesia* which make up a

The unique flora and fauna of the Galapagos Islands make the archipelago a treasure house for naturalists. Recently the government of Ecuador, administrating the group, has converted all the uninhabited islands and parts of the inhabited ones into nature reserves. There is also a biological station for scientific research.

*Facing page:* Among the fascinating animals found along the rocky coasts of the Galapagos are albatrosses (*above left*), penguins (*above right*) and sea-lions (*below*). Like most of the animals of the archipelago, they show little fear of humans.

region of dry forest, except in certain spots where winds drive clouds against the mountainsides.

Above the 1,000-foot mark these trees are replaced by guavas (*Psidium galapageum*) which constitute the so-called brown zone, the upper limit of which is at about 1,800 feet. From that point up to approximately 2,000 feet climatic conditions are unsuitable for tree growth and the most typical forms of vegetation are shrubs, especially those of the genus *Micania*.

Higher up, above 2,100 feet, the temperature is so low and the winds so dry that there is little save grass and ferns.

## Marine animals of the Galapagos

Having briefly described the successive vegetational zones that are likely to be encountered, although naturally with local variations, on the Galapagos, it is time to turn to the wildlife of these remarkable islands, going back to the final stages of the sea-voyage, prior to dropping anchor in one of the bays of black lava, off a shore lashed by foaming waves.

Long before setting foot on an island we will probably have had a foretaste of the wonders to come. Towards the end of the journey we will perhaps have seen flocks of brown pelicans and frigate-birds. These two species sometimes clash, for the frigate-bird is unable to dive beneath the surface and will often chase a pelican with a fish in its beak, trying to make it drop the prey into the sea. We may also have been fortunate enough to disturb a school of whales. Nearer to the islands, as the ship edges through the Bolivar Strait between Fernandina and Isabela, a careful scrutiny through binoculars may reveal a scene which is more reminiscent of the Antarctic than the equator. For the birds sunning themselves on the rocks or swimming in the sea appear to be penguins. This is no mistake. They are in fact small Galapagos penguins (*Spheniscus mendiculus*) and their presence here can be explained by the passage of the cold Humboldt current. By following this 'river' in the middle of the ocean the ancestors of these penguins travelled to the Galapagos from the southern tip of South America. Here, in complete isolation, their descendants evolved and eventually formed a new species.

Mingled with these penguins we would also be likely to see flightless cormorants (*Nannopterum harrisi*), birds with atrophied wings, standing bolt upright on a rock in the sea or plunging into the waves for fish.

Penguins were not the only 'castaways' to have made use of the Humboldt current to reach the lost islands of the Pacific. Seals which started out from the coasts of South America were the ancestors of a local subspecies, the Galapagos fur seal (*Arctocephalus australis galapagoensis*); and it was by similar seaways that the Californian sea-lion (*Zalophus californianus*) reached the islands, eventually giving rise to the Galapagos sea-lion (*Zalophus californianus wollebaeki*).

In the Galapagos Islands the fur seal has been subjected to merciless exploitation and today there are only about 1,000 left, concentrated on the islands of Fernandina, Isabela and

San Salvador. The situation of the Galapagos sea-lion is somewhat happier because its fur is of no commercial value and the animal is therefore not much hunted. They are still to be found in large numbers, especially on Española. This island is also the only nesting site in the world of the Galapagos albatross (*Diomedea irroratea*), of which there are estimated to be only a couple of thousand surviving pairs.

## Iguanas of sea and land

It is often said that visiting the Galapagos Islands is like taking a journey back into prehistory, to the age when the earth was dominated by giant reptiles. This is a bit of an over-simplification, but accurate in one sense. The fauna of the islands consists almost exclusively of reptiles; but these animals are not relics of the Mesozoic age that have survived to this day. It is simply by chance that reptiles, rather than mammals, somehow managed to make the long journey from a distant river estuary in America to the Galapagos, possibly by means of floating tree trunks and branches.

Nevertheless it is not difficult to imagine that one has wandered back to the beginnings of time as dawn reveals hordes of huge, bizarre-looking lizards emerging from their nightly shelters among the rock clefts. Some of them measure more than 5 feet long. Notable characteristics include a flat snout, bow legs, a large tail flattened from side to side, and an erectile crest of spines along the neck and back. Their colour ranges from dark grey to black, but some individuals have red patches on the body and forelegs, with the dorsal crest highlighted in green. These lizards are marine iguanas (*Amblyrhynchus cristatus*) and they are found only on this archipelago.

As soon as they leave the sheltered places where they have spent the night the iguanas expose their bodies to the sun. It is a remarkable experience to see them basking on the warm rocks as large red crabs scuttle all over them to feed on the parasites lodged in their skin. But the tolerance which the marine iguanas show towards crabs is extended to other creatures, man included. One can walk up to them, touch them, grab them by the tail, even pick them up, without their giving the least hint of being frightened or showing any desire to escape. This astonishing confidence and trust, common to all the animals of the Galapagos, is of course the result of a long period of evolution free from natural enemies.

The instinct of fear and its corollary, flight, are normally indispensable for survival. Many species would long since have been destroyed by carnivores if they had not had recourse to self-preservation. But these iguanas have experienced few predators, apart from the odd shark, and have therefore not acquired the reflex instinct of fleeing to safety.

As the tide goes out and the rocks where the iguanas have been sunning themselves are surrounded by belts of seaweed, the huge reptiles move off in the direction of the sea, plunging into the waves in search of food which consists exclusively of seaweeds. When the water is especially rough, as in the places

Marine iguana
(*Amblyrhynchus cristatus*)

Land iguana
(*Conolophus subcristatus*)

**MARINE IGUANA**
(*Amblyrhynchus cristatus*)

Class: Reptilia
Order: Squamata
Family: Iguanidae
Total length: up to 67 inches (170 cm)
Diet: seaweeds
Number of eggs: 2-3
Incubation: 110 days

Massive body with flattened snout and long dorsal crest from neck to tail. The tail is flattened to facilitate swimming. Colour dark grey or black, some individuals with reddish patches on body and forefeet and with green crest.

**LAND IGUANAS**
(*Conolophus* spp.)

Class: Reptilia
Order: Squamata
Family: Iguanidae
Total length: up to 43 inches (110 cm)
Diet: flowers, fruit, leaves, shoots

Smaller but heavier than marine iguanas, lighter in colour but with a similar dorsal crest. Two Galapagos species, *Conolophus pallidus*, exclusive to Santa Fé, and *Conolophus subcristatus*, more widely distributed.

*Facing page:* In some areas the rocks seem to be covered with marine iguanas, over which scamper large red crabs.

Unlike the marine iguanas which live along the coasts, feeding on seaweeds, the land iguanas inhabit the dry inland zones. Their food consists of many kinds of vegetation but principally of the flowers and fruits of cacti.

where strong currents threaten to carry them out to the open sea, the iguanas grip the rocks with their long claws. Then at mid-day, almost as if responding to a signal, they return to the shore, appetites satisfied, and go to sleep in a shady spot.

In the breeding season the males stake out small tracts of territory, squabbling noisily among themselves but limiting their fights to ritual mouth openings and head swayings. Those that come off second best and beat a retreat are invariably individuals least suited for a reproductive role. Once the triumphant males have established themselves they are free to mate with any passing female, more or less at random. Courtship simply consists of pursuit, the male attempting to seize the female by the scruff of the neck. If and when he succeeds the two reptiles mate.

After coupling, the fertilised females assemble on selected beaches where each digs a tunnel, about one foot long, in which two or three eggs are subsequently laid, covered with sand and promptly abandoned. The babies hatch 110 days later.

If one strikes inland from the coast through the lowland zone of cacti, there can be no missing other lizards, similar to the marine iguanas in general appearance, which are content to feed on the flowers and fruits of the prickly pear as well as on the leaves and shoots of other plants. These are land iguanas, of which there are two species, *Conolophus subcristatus* and *Conolophus pallidus*. The former used to be found on the islands of Fernandina, Isabela, San Salvador, Santa Cruz and three nearby islets. The latter inhabits Santa Fé.

Slightly smaller but somewhat heavier than their marine counterparts, the land iguanas are also more shy.

Although they feed on many types of vegetation, most plant species have only a brief flowering season so that for the greater part of the year prickly pear and other thorny cacti provide the reptiles' principal fare. The cactus spines do not appear to bother them and are expelled with the feces after travelling harmlessly through the intestine.

Because of their feeding habits the land iguanas of the Galapagos never go hungry, and the dense underbrush provides splendid protection for the young which during their early weeks are prey to the Galapagos hawk (*Buteo galapagoensis*).

The appearance of man among the islands of the archipelago brought tragedy for large numbers of these reptiles which, until then, had lived peacefully for thousands of years. There were both direct and indirect repercussions. Being hunted for their skins was bad enough, leading to a dramatic reduction of numbers, but even more catastrophic was the effect of introduced animal species from foreign lands. The absence of animals, particularly mammals capable of providing fresh meat, forced the 17th-century pirates who set up bases in the islands to let loose domestic goats. As the population of goats increased they ate more and more vegetation, depriving the lizards of food and exposing the babies to raptors. When the Spaniards attacked the islands they tried to eliminate the goats by bringing in dogs and thus force out the buccaneers who continually raided their shipping; but the dogs found the reptiles much easier prey and once more the iguanas were decimated.

Today the largest communities of land iguanas are on Fernandina, which has been the least affected by human settlement, and on Isabela. On San Salvador, where they were still abundant when Darwin visited the island in the *Beagle,* they have not been seen since the beginning of the present century; on Santa Cruz their numbers are small. As for the three adjacent islets, two still provide the iguanas with a refuge but the third has recently been depopulated. The first islet harbours a tiny community, the second, now free of goats, has witnessed an iguana revival. This islet is only 100 yards long and 150 yards wide, yet about one hundred iguanas have been counted here.

On Santa Fé the population was reasonably high in 1957 but because of sparse plant cover which makes hunting conditions ideal for hawks, the numbers have now declined.

## The giant Galapagos tortoises

The first Spaniards who visited the archipelago were particularly impressed by their discovery of giant tortoises, similar in appearance to the much smaller freshwater tortoises which they remembered from the rivers of their homeland. It seemed appropriate, therefore, that they should apply their word for 'tortoise' to the islands they had brought to light.

The size of these tortoises (*Geochelone elephantopus*) is astonishing, for they measure up to 5 feet in length and may weigh 550 lb. How could such immense animals possibly have encompassed the 600 miles separating the islands from the mainland of South America? Detailed study on this subject has convinced naturalists that these tortoises are capable of floating in sea water for a considerable time. We know that the South American ancestors of these reptiles were nothing like so large during the

The Galapagos hawk feeds on the young of land iguanas and tortoises. It has been aided in its hunting by imported goats which have fed so abundantly on plants that they have exposed these reptiles to the view of this powerful raptor.

**GALAPAGOS TORTOISE**
(*Geochelone elephantopus*)

Class: Reptilia
Order: Testudines
Family: Testudinidae
Total length: up to 60 inches (152 cm)
Weight: up to 550 lb (250 kg)
Diet: vegetation
Number of eggs: 6–11

Largest land tortoise in the world, found only in the Galapagos. Numerous subspecies, some rare, distinguished mainly by different types of shell. Recent studies suggest that they are not as long-lived as was once believed.

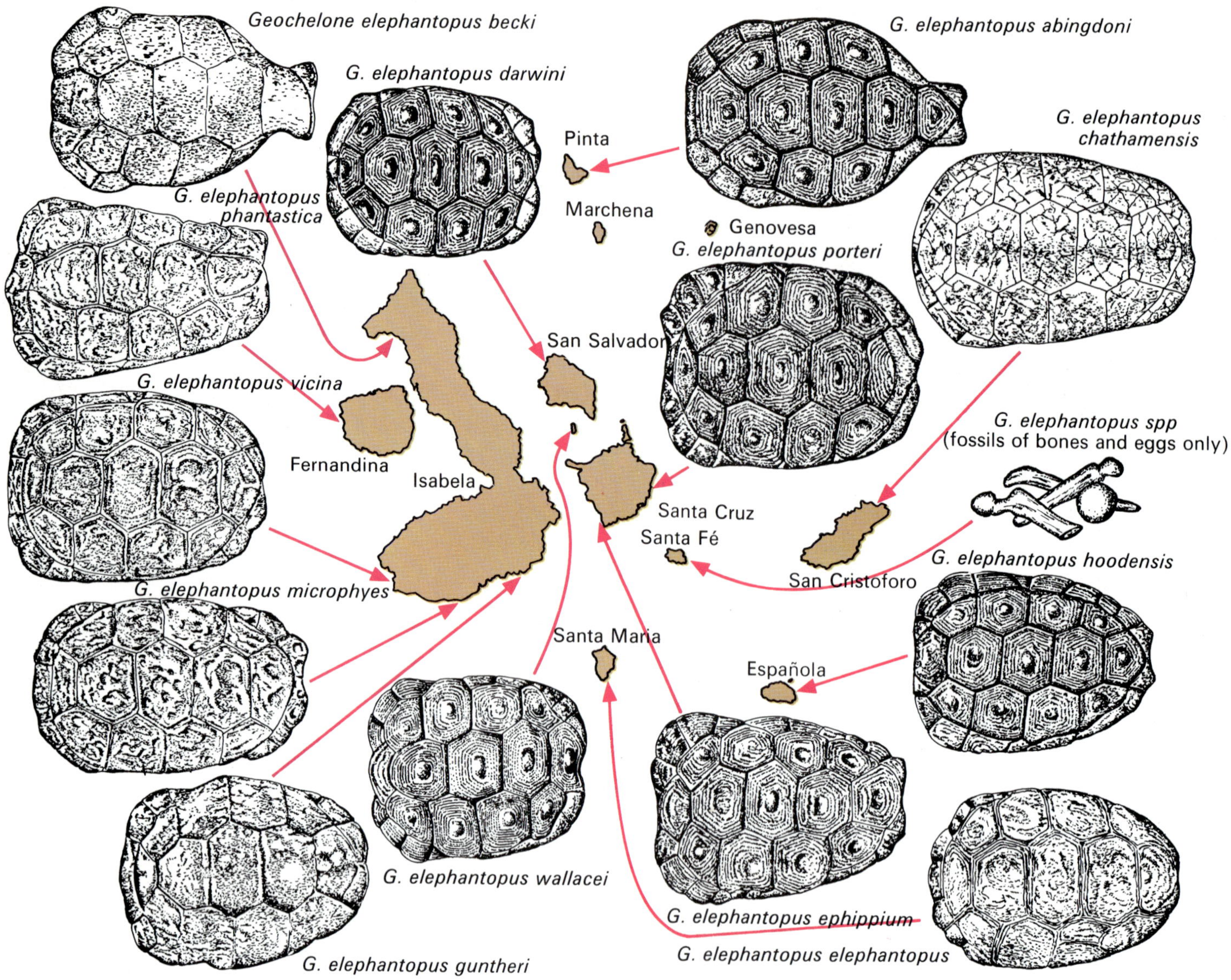

Miocene epoch, which is approximately when the Galapagos Islands were first colonised by animals, so that this was probably how they travelled. But we cannot discount the possibility that for part or indeed all of their voyage they may have taken advantage of natural rafts, including tree trunks.

Even more perplexing than the way in which the tortoises originally got to the Galapagos is the amazing fact that many of the islands contain their own type of tortoise. Since it stretches reason too far to suppose that each distinct form could have made the long sea journey and landed up on a different shore, the only conclusion must be that separate development took place here in the Galapagos, after the arrival of the ancestral type. Most specialists agree that the giant tortoises of the Galapagos Islands are not separate species but different subspecies of *Geochelone elephantopus*. It is a wonderful example of diversification of a species stemming from common ancestral stock, occurring purely as a result of geographical isolation.

Most of the races of giant tortoises inhabit separate islands and there has been no hybridisation. Even in places where several forms live together, as on Isabela, cross-breeding is impossible because each community lives on the slopes of one or other of the island's five volcanoes, the distances between them being more than their normal journeyings.

The giant tortoises of the Galapagos are herbivorous, feeding on all kinds of plants, including cacti. As was the case with the land iguanas, the introduction of goats had profound repercussions, for the tortoises were far too slow and clumsy to compete with them for food. Later the dogs inflicted heavy losses on them, particularly the young, while rats and pigs devoured their eggs.

## Massacre and rescue

Ever since the Galapagos Islands were discovered, their flora and fauna have been systematically destroyed. In the case of the giant tortoises hunting assumed the proportions of a massacre. Their flesh was particularly prized and special expeditions were mounted to capture and kill these huge, defenceless creatures. According to the biologist C. H. Townsend, who examined the logbooks of 105 American whaling vessels for the years 1811-1846, some 15,000 tortoises must have been wiped out by the crews of these ships. The average is just over 140 animals per ship and if one multiplies this figure by the estimated total number of ships visiting the islands during these years, the unbelievable scope of the slaughter can be appreciated. In the opinion of the same author the number of giant tortoises killed for their flesh and shells after the islands were discovered must have been in the region of ten million!

To save the species the government of Ecuador, to whom the islands belong, passed strict laws in 1934 to protect the tortoises and other species of flora and fauna, but these were not properly enforced. In 1957 UNESCO sent out a team of specialists. In their report they recommended setting up nature reserves and a research centre; and in the following year Jean

## HISTORY OF THE TORTOISES OF THE GALAPAGOS ISLANDS
### (IUCN Red Data Book)

| History | Reasons for present situation |
|---|---|
| **Pinta (Abingdon Island)** | |
| Abundant in 1882 | Excessive hunting |
| Very rare in 1900 | Introduction of goats |
| Extremely scarce in 1962 | between 1957 and 1963 |
| Extinct in 1964 | |
| **San Salvador (James Island)** | |
| Abundant in 1812 | Excessive hunting |
| Very rare in 1870 | |
| Still surviving, but numbers not known, in 1966 | |
| **Rabida** | |
| Extinct; once abundant | |
| **Pinzón (Duncan Island)** | |
| Abundant in 1905 | Introduction of rats |
| Very rare in 1962 | which devoured young |
| 140 individuals in 1964, 90 in 1967 | |
| **Santa Cruz (Indefatigable Island)** | |
| Abundant in 1870 | Establishment of reserve |
| About 2,000 individuals today | |
| **Santa Fé (Barrington Island)** | |
| Present in 1853, now extinct | Excessive hunting |
| **San Cristoforo** | |
| Abundant in 1813 | Excessive hunting |
| Rare in 1863 | Sale of young as souvenirs to soldiers of army base around 1940 |
| Very rare in 1905 | |
| Reported extinct in 1960 | |
| Extremely rare today | |
| **Española (Hood Island)** | |
| Abundant in 1831 | Excessive hunting |
| Scarce in 1853 | Introduction of goats |
| Very rare in 1906 | |
| Extremely rare today | |
| **Floreana** | |
| Very abundant in 1812 | Excessive hunting |
| Very scarce in 1840 | |
| Extinct around 1850 | |
| **Fernandina (Narborough Island)** | |
| Surviving today but numbers unknown | Difficulty of access hinders hunting |
| **Isabela (Albemarle Island)** | |
| Abundant in 1860 | |
| All island's subspecies still surviving | |

*Facing page*: Geographical isolation has led to the diversification of the islands' giant tortoises, distinguishable by their shell structure.

*Following pages*: Group of marine iguanas.

The different genera of Darwin's finches exhibit variations of diet and habitat. Thus the finches of the genus *Geospiza*, one of which is shown in the photograph, are ground birds, whereas those of the genus *Certhidea* are tree-dwellers. Others occupy intermediate positions.

---

**DARWIN'S FINCHES**

Class: Aves
Order: Passeriformes
Family: Fringillidae

Belonging to subfamily Geospizinae, consisting of six genera and fourteen species. Some are seed-eaters, with terrestrial habits, others are insectivorous and live in trees. In between are species that have a more varied diet, including seeds, insects, fruit and flowers. The shape of the bill is the chief distinguishing feature between the species.

---

Dorst was detailed to complete studies for the establishment of a biological station. The outcome was the Charles Darwin Foundation for the Galapagos, the President of which is Sir Julian Huxley, its directors including distinguished scientists from various nations. At the same time the government of Ecuador tightened up their legislation. Today all the inhabited islands of the archipelago and the western part of Isabela are nature reserves and the giant tortoises have been saved.

## The miracle of Darwin's finches

The visitor to the Galapagos cannot help being impressed, first and foremost, by the strange iguanas and giant tortoises; and it may be some time before he becomes aware of the masses of small birds, about the size of sparrows, which hop about on the ground or flit among the branches. This is perhaps not surprising for these little Darwin's finches (named after the famous naturalist who first took notice of them and recorded their diversity of form and behaviour) have fairly nondescript plumage. Their scientific interest has nothing to do with brilliance of colour and pattern but is due to the manner in which they exemplify the evolution of new species. Whereas in the case of the giant tortoises it was geographical isolation which determined the diversification of the species, the Darwin's finches (recognised as distinct species rather than subspecies) have gone their separate ways as a result of intraspecific rivalry and competition. These small birds have been compelled to seek new habitats and in so doing have undergone a series of adaptations in their struggle for survival.

Darwin's finches are thought to be the descendants of South American finches which are now extinct and which were probably driven by strong winds to the Galapagos Islands. Like other Fringillidae these new arrivals fed on vegetation and it is likely that such food consisted in the main of seeds collected from the ground. Their habits were therefore essentially terrestrial. But the various islands offered much more than seeds in the way of food, and with relatively few other bird species to bother about the finches were mainly competing with one another. Since there were soon too many finches trying to feed on seeds, some of them modified their diet to include flowers, fruit and even insects. This led in due course to the separation of finch communities and their assembly in small groups which later developed into different species.

There are today three recognised genera and thirteen species of Darwin's finches, all native to the Galapagos apart from one which inhabits the Cocos Islands. They can sometimes be identified by size and colour of plumage but the easiest method of recognition is the shape of the beak. They range between two extremes. The finches of the genus *Geospiza* are still seed-eaters and their bill is short and thick, whereas that of the genus *Certhidea,* which are insectivores, is long and slender. The former spend most of their time on the ground, the latter in trees. The finches of the four other genera eat seeds, fruit, flowers and insects in varying proportions, some of them being better adapted to life on the ground, others to life in trees.

Each genus comprises one or more species, and the differences between the species shows to what extent each has adapted to one type of life rather than another. Thus among the seed-eating finches of the genus *Geospiza* there are two closely related species, the cactus ground finch (*Geospiza scandens*) and the large cactus ground finch (*Geospiza conirostris*), which do not compete because they live on different islands. Three others, the large ground finch (*Geospiza magnirostris*), the medium ground finch (*Geospiza fortis*) and the small ground finch (*Geospiza fuliginosa*) live together in some places; but the bill of the first is larger and stronger than that of the second, which is in turn bigger than that of the third. Each feeds on different types of food and in areas where they are found together each keeps to a separate small zone, so that rivalry is reduced to the minimum.

One species is of particular interest, namely the woodpecker finch (*Geospiza fuliginosa*) live together in some places; but the bill of the first is larger and stronger than that of the second, the Egyptian vulture, which smashes ostrich eggs by dropping rocks and stones on them. The Galapagos woodpecker finch feeds on insects lodged in tree bark and is thus the counterpart of the woodpeckers of the Holarctic region. Unlike the latter, however, it does not have a long, protractile tongue. Sometimes it can get at its victim with the beak alone, but if a grub is too deeply embedded in the bark the bird picks up a cactus spine in its bill to scoop it out. The finch does not do this in a systematic way but only when the need arises. Yet when the spine is either too long or too short the bird will readjust its position in the beak so that it can effectively function as a tool, a genuine example of intelligence. On one occasion a naturalist watched a woodpecker finch trying to manipulate a spine with a forked tip. Rather than abandon it the bird turned it around and used the opposite end.

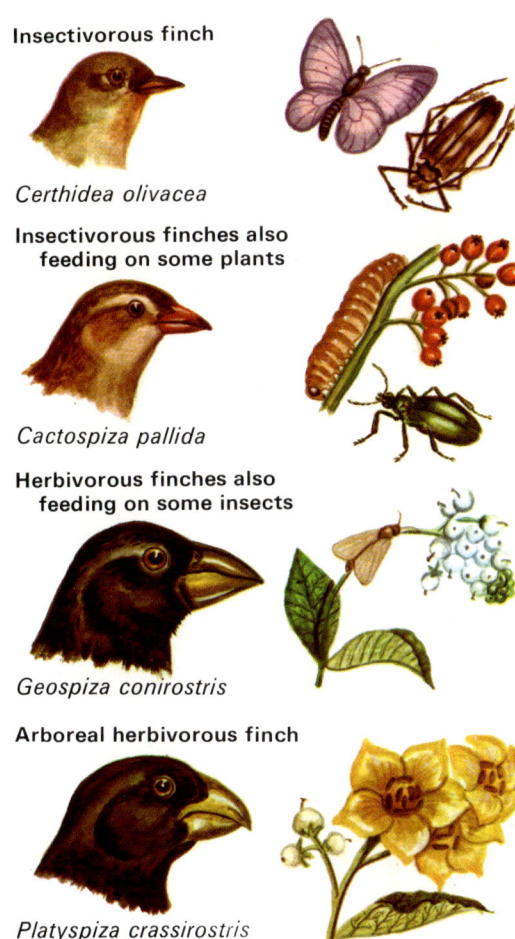

**Insectivorous finch**
*Certhidea olivacea*

**Insectivorous finches also feeding on some plants**
*Cactospiza pallida*

**Herbivorous finches also feeding on some insects**
*Geospiza conirostris*

**Arboreal herbivorous finch**
*Platyspiza crassirostris*

Darwin's finches do not compete for food. Those that are not exclusively insect-eaters or plant-eaters combine both forms of food in varying measure.

The woodpecker finch resembles the true woodpeckers of other regions in that it feeds on grubs lodged in tree bark. Lacking a long protractile tongue, however, it often uses a cactus spine to scoop out its victim.

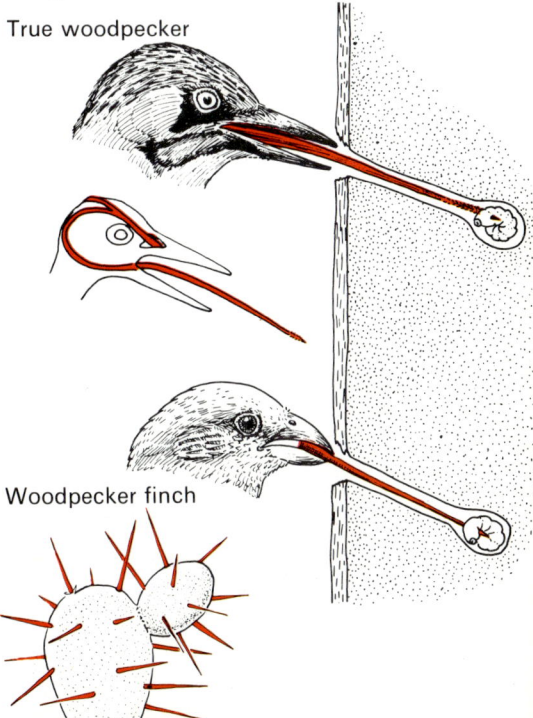

True woodpecker

Woodpecker finch

# CHAPTER 107

# Paradise of the Pacific isles

The largest ocean in the world, the Pacific, stretches from the east coast of Asia to the western shores of the Americas. It is dotted with innumerable islands and islets. Some of them record human habitation going back two thousand years, but only over the last four centuries have the legendary South Sea islands been discovered by Europeans.

The peace and security of these island paradises were to be shattered by the arrival of settlers from Europe. Densely covered with greenery and enjoying the pleasantest of climates, the islands were havens of wildlife where no dangerous predators roamed. But the white man disturbed the fine balance of ecosystems, many animal species were wiped out and the native human populations succumbed to hitherto unknown diseases introduced from abroad. Ancient customs and life patterns were debased or demolished by the waves of Western invasion.

Nevertheless these delightful islands with their splendid scenery and friendly people continue to attract tourists in ever increasing numbers; and to the naturalist they offer a living laboratory of plants and animals, some clearly of ancient origin, others obviously related to familiar continental species but which have evolved distinct forms and behaviour patterns that are to be found nowhere else.

## The wildlife of Oceania

Looking at a map of the vast Pacific and the scattered islands of Oceania it is difficult to imagine any possible relationship among the various animal populations. Yet these islands are a biological entity. Their inhabitants are related to one another and the manner in which they arrived at their different desti-

*Facing page:* The numberless islands of the Pacific Ocean, although no longer the peaceful havens of bygone days, still harbour many fascinating plants and animals and may be likened to a huge biological laboratory of wildlife.

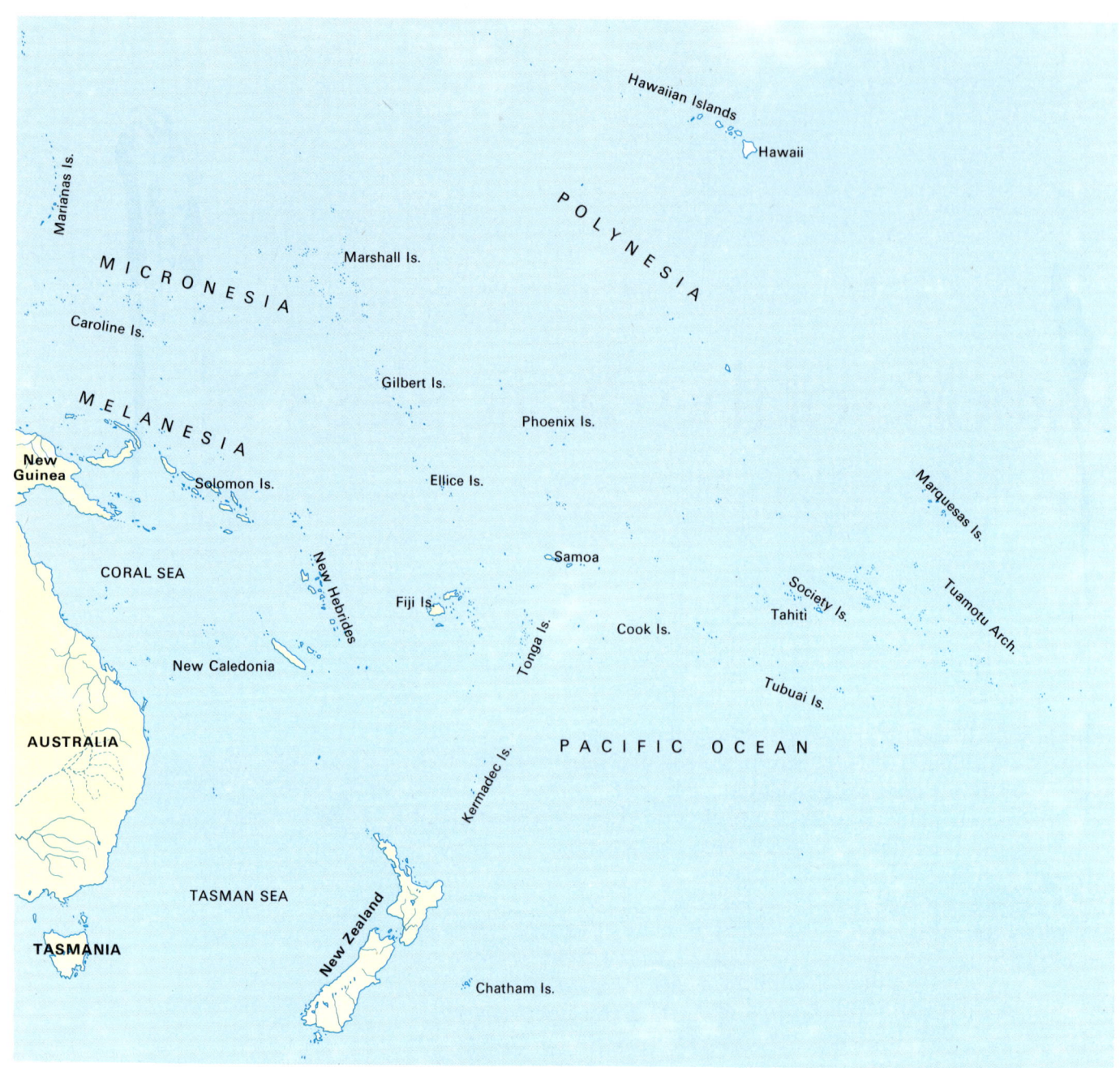

Oceania is a vast biogeographical region containing innumerable islands, the animal inhabitants of which are linked by a consistent general pattern of migration and colonisation.

nations is part of a consistent pattern of migration. So it is quite logical to study the distribution of animals in the Pacific islands as one would the wildlife of any other compact biogeographical region. The ocean barriers physically separating them are in this sense unimportant.

Because the islands are so far apart, however, the area as a whole is poor in zoological species. Almost all the animals have links with those that populate Asia and Australia, apart from a few rare species that originated in America. It now seems certain that there was a great wave of migration across the Pacific from west to east, with some species managing to travel farther than others and reaching more distant shores. Moving attention eastwards from Australia and New Guinea, it is noticeable that there is a steady drop in the number of animal groups but a proportionate increase in the number of endemic species, another consequence of geographical seclusion. Thus, according to statistics produced by Jean Dorst, there are in the islands lying close to Asia 503 species of birds, of which 12.7 per cent are endemic. On the western and central Pacific islands there are 255, and 27 per cent of these are native species. The islands of the eastern Pacific accommodate only 42 bird species but of these 33 are endemic, equivalent to approximately 78.6 per cent.

Not only are there comparatively few species of animals but some groups are completely missing. Thus in the majority of islands there are neither amphibians (which are highly sensitive to salt water) nor reptiles. Mammals, which are poor travellers, can only cross short stretches of open sea; but there are some marsupials—the cuscus and pademelons of the genus *Thylogale*—on New Britain and New Ireland. The former also reached the Solomon Islands. But since flying animals have no such problems there are numerous birds and bats in Oceania.

Most of the bats of the Pacific islands are fruit-eating species, notably the large flying foxes of the genus *Pteropus*, which are in fact widely distributed from Madagascar to Samoa. Among the numerous bats of the Solomons are those of the genera *Rousettus* and *Dobsonia*, and the strange bats of the genus *Nyctenema* with their long tubular nostrils. There are also a number of species belonging to the genera *Melonycteris* and *Nesonycteris*, with reduced teeth and a long protractile tongue for sipping pollen and nectar.

The insectivorous bats travelled by more or less the same paths as the fruit-eating bats. Certain members of the family

Among the many birds of Oceania are the little pied cormorant (*left*) and the cardinal honey-eater (*right*).

Emballonuridae (tomb and ghost bats) travelled as far as Samoa. This is the limit to the distribution of the Hipposideridae (leaf-nosed bats) and Vespertilionidae, the latter represented by the bats of the genus *Myotis*, but beyond the bounds of the Rhinolophidae. All these bats originated in Malaysia and New Guinea; but those genera that colonised Hawaii and the Galapagos are descended from the American genus *Lasiurus*.

## The fascinating Pacific birds

Of all the animals that nowadays inhabit the Pacific islands the most interesting are undoubtedly the birds. They have been the most successful of the ocean voyagers and because of their singularities they offer ornithologists a most rewarding and exciting field of study.

The evolution of certain groups, sometimes of ancient stock, in conditions of complete isolation, has led to an astonishing variety of forms. A classic example of such diversification is that of the grey-headed thrush (*Turdus poliocephalus*) which is represented by forty subspecies distributed through Malaysia and Oceania, differentiated by the colour of their plumage. The variation is most marked in Fiji where there is an endemic form on each island. On Taveuni the subspecies is black with a grey crest (*Turdus poliocephalus tempesti*); on Kandavu it is black with a red crest (*T. p. bicolor*); on Viti Levu, grey with a red belly (*T. p. layardi*); on Vanua Levu the feathers of the abdomen are fringed with russet (*T. p. vitensis*); and on Ngau there is a completely black variety (*T. p. hades*).

Furthermore, the absence of ground predators has, in certain cases, resulted in the loss of flying capacity, as among the rails of the family Rallidae, which, in spite of normally sedentary habits, have settled on several islands, including Fiji, Samoa and Tahiti. These too are found in diverse forms, according to the particular locality.

Other groups of birds that settled in the islands gave rise to completely new species. Oceania is, for example, a paradise for doves and pigeons (Columbiformes). The imperial pigeons (Duculinae), one of the groups of fruit pigeons, are distributed from Malaysia to the Marquesas and Tuamotu archipelagos.

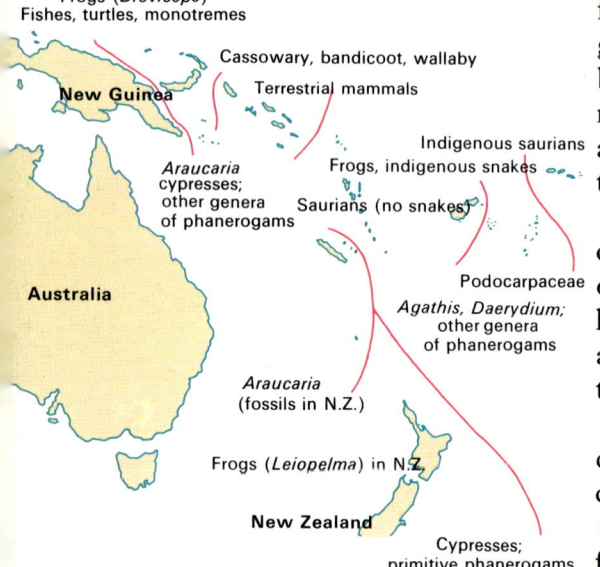

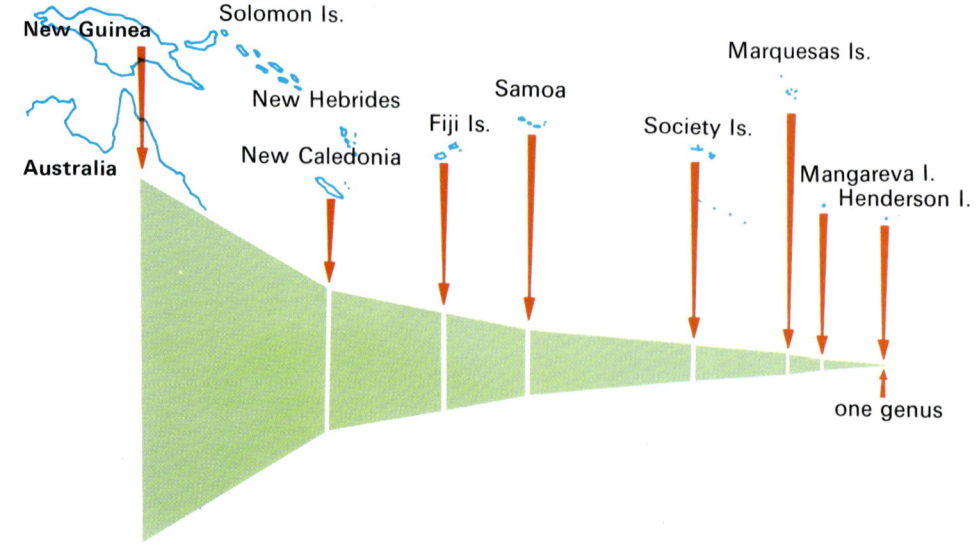

The majority of plants and animals found in the Pacific islands came from the west. The more mobile the group the farther it was able to travel. Not surprisingly, therefore, the numbers of genera decrease steadily as one moves eastwards from Australia and New Guinea. The diagram on the right shows this phenomenon in the case of beetles of the family Cryptorhynchidae.

## RELATIVE CAPACITY OF LONG-DISTANCE DISPERSAL OF ANIMALS
(according to S. Carlquist)

| Animal group | Distance | Examples |
| --- | --- | --- |
| Freshwater fishes | Able to travel a few miles in salt water | |
| Large mammals | At most 25 miles | Exceptions are semi-aquatic mammals (hippopotamus on Madagascar) |
| Small mammals (apart from rodents) | 200 miles | Probably Viverridae and Insectivora on Madagascar |
| Rodents | 500 miles or more | Galapagos Islands |
| Amphibians | 500–1,000 miles | Seychelles, New Zealand |
| Freshwater turtles | 200 miles | Madagascar |
| Land tortoises | 500 miles or more | Galapagos Islands |
| Snakes | 500 miles or more | Galapagos Islands |
| Lizards | 1,000 miles | New Zealand (tuatara), probably farther for Gekkonidae |
| Bats | 2,000 miles | From N. America to Hawaii |
| Ground birds | 2,000 miles or more | From N. America to Hawaii, from S. America to Tristán da Cunha |
| Ground molluscs | Over 2,000 miles | From Polynesia to Juan Fernández |
| Insects and spiders | Over 2,000 miles | |

These birds are remarkable for the structure of the bill. The lower mandible can be thrust down and forwards, forming an angle of almost 180°, in order to cope with varieties of large fruit, somewhat in the manner of snakes. On Savai'i, one of the Samoan islands, there is another strange pigeon, the tooth-billed pigeon (*Didunculus strigirostris*).

## Colonisation of an island

A naturalist will probably point out that animal inhabitants of islands are closely related to those of nearby continents, the main differences among them usually being nothing more than the result of the long isolation of the former species. Thus in very ancient times the ancestors of the present-day island residents must have come from adjoining land masses. When one realises that the geography of this planet is far from static and that it has undergone tremendous upheavals in the course of geological ages, the obvious conclusion is that those regions that are now islands were formerly attached to continents, or that they were part of a larger land mass that has since been submerged. But although this may be accurate in certain circumstances it by no means explains the distribution of all living creatures, as was thought to be the case when science admitted only one solution, that of the continental bridge.

Today scientists have at their disposal a mass of information derived from numerous sources which has led them to conclude that although some land links undoubtedly did exist and could help to explain certain aspects of global distribution, such a theory is valid only for continental islands. The oceanic islands, on the other hand, whether of volcanic or coral origin, have at no time been attached to any continent, which means to say that the various animals colonising them must have arrived by some other means. One must bear in mind, for example, that such islands, particularly coral islands, can spring up and perhaps vanish very quickly, even in the course of a single day. In other words, lands that have now disappeared under the

The earliest human settlers of the Pacific islands discovered trees which bore large, starchy, highly nourishing fruit. This so-called breadfruit is still the staple fare of the South Pacific islanders.

Map of the Krakatau (Krakatoa) islands.

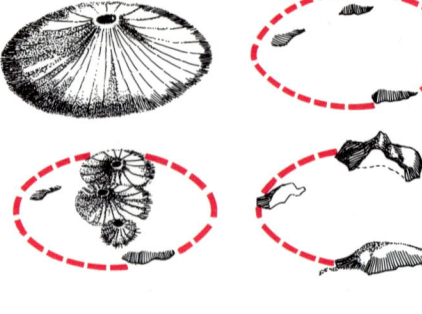

Four phases in the eruption of Krakatoa.

waves may at one time or another have formed stepping stones from island to island. Given all the available information, experts in biogeography are today convinced that the installation of living creatures on islands does not necessarily imply a former link with a continent.

Dispersal over immense distances can be achieved in several different ways. The easiest to imagine is simple transport by sea, but paradoxically this method has its limitations. Although certain plants can be distributed in this fashion, especially those with seeds resistant to salt water (as is the case with the coconut), the majority of animals cannot travel in this manner; even fishes accustomed to shallow coastal waters are unable to survive very long in the open sea. But the ease with which plants can be dispersed is a disadvantage as compared with animals, in which new species are evolved by adaptive radiation from a small number of ancestral types. The continuous arrival of new seeds, all possessing the same genetic structure as the original population, reinforces the original genotype, and thereby rules out the possibility of the emergence of new forms.

The second method of island colonisation is more surprising yet clearly of greater significance than was at one time believed. This is by means of floating objects on which terrestrial animals can be carried for considerable distances.

Methods of plant and animal diffusion over long distances.

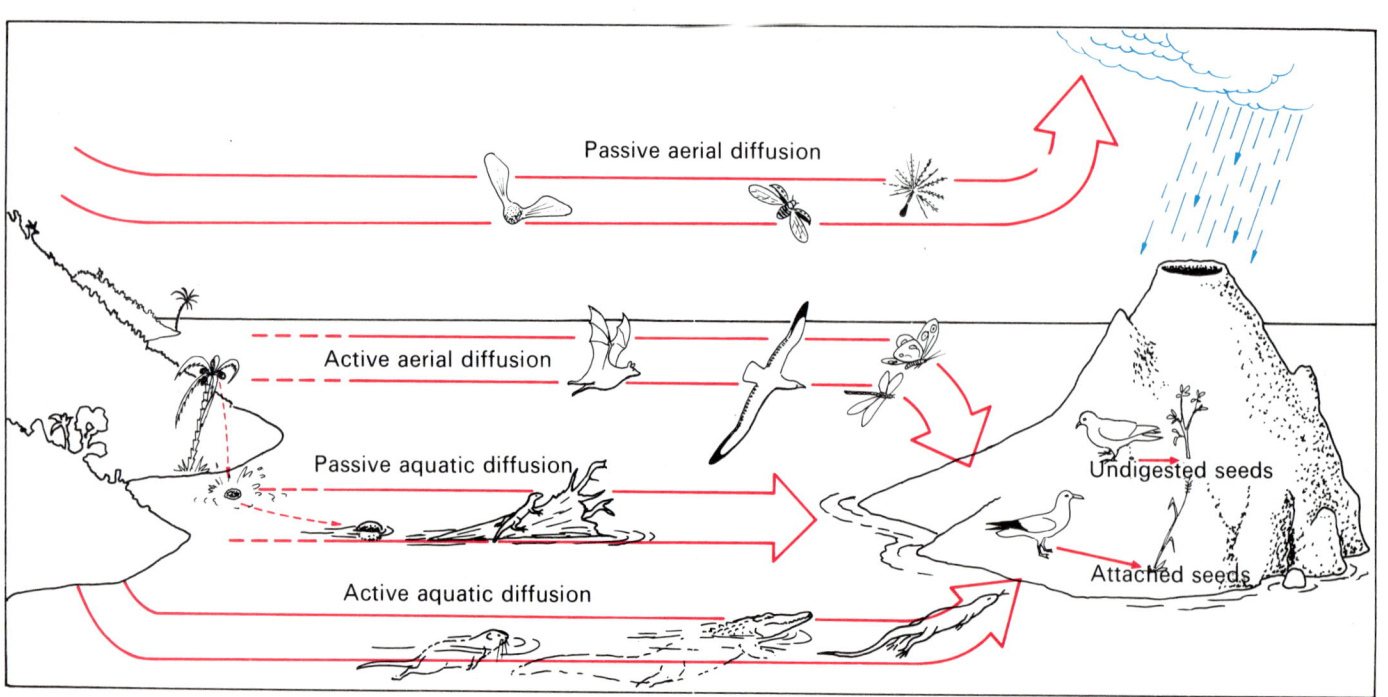

Perhaps the most surprising method of animal dispersal is by means of floating objects such as tree trunks. Those coloured here in red are good, adaptable travellers, those in green moderately good and those in brown poor, unable to survive very long.

There can be no doubt that this is a frequent occurrence, for a mass of eye-witness reports have verified it. All the world's streams and rivers carry down to the sea a miscellany of floating objects—branches, tree trunks and bundles of vegetation—some of which are stable and sizeable enough to provide a means of transport for involuntary passengers; and violent storms along the coasts are also capable of removing young trees, roots and all, and driving them far out to sea.

The distances such 'rafts' can travel depends, of course, on the weather. Currents which are hardly noticeable when it is calm may be transformed into unbelievably powerful submarine torrents under the stress of storms. In such circumstances a large tree trunk is likely to be swept along for many miles, bobbing up and down with only a few inches below the surface. The kinds of animals which may be transported passively in this way obviously vary enormously, some being better able than others to withstand the hardships of such a voyage. Thus mammals are, in principle, poor travellers and are unlikely to survive for long. But lizards and snakes, with physiological mechanisms that assure survival for considerable periods in the absence of food or fresh water, can cover vast distances. Land molluscs are also good sea travellers.

A third important factor in plant and animal dispersal is aerial transport. The lighter a floating particle, the farther it will be carried by the wind. Here too, diffusion will be stimulated by special circumstances. A hurricane, for example, may sweep heavy particles along at fantastic speed. Furthermore, turbulent air currents in the upper layers of the atmosphere also possess tremendous carrying power.

In any event, the possibility of such seeds coming down on an island is not just a matter of chance. When an air current collides with a mountain range and is forced upward, the water vapour content condenses; seeds, spores and other heavy particles floating in the air are focal points of condensation so that they are enveloped in droplets of water. This increases their weight and they fall to earth. Islands with high volcanoes therefore collect enormous quantities of atmospheric particles. Aerial transportation is also far easier for seeds with 'wings' that function as miniature parachutes.

As far as animals are concerned insects are naturally good aerial travellers and may be carried huge distances by the wind. To prove this point the Hawaiian entomologist J. Linsey Gressitt conducted an interesting experiment, filtering the air (by means of fine nets dragged by ships and aeroplanes) over a selected zone in the South Pacific. According to his calculations he filtered approximately 19,000 cubic yards of air, capturing 1,065 insects.

Much information about the ways in which plants and animals colonise islands has been obtained from detailed study of the flora and fauna of the Krakatoa islands after the volcanic eruption of 1883. The diagram on the opposite page shows in four successive stages the probable aspect of the volcano before and after an eruption in primitive times, and before and after the eruption of 1883. The graph below shows the various methods and frequency of plant dispersal (by sea, by air, by animals and by man) during the period 1883–1934, the numbers referring to new species appearing on the islands.

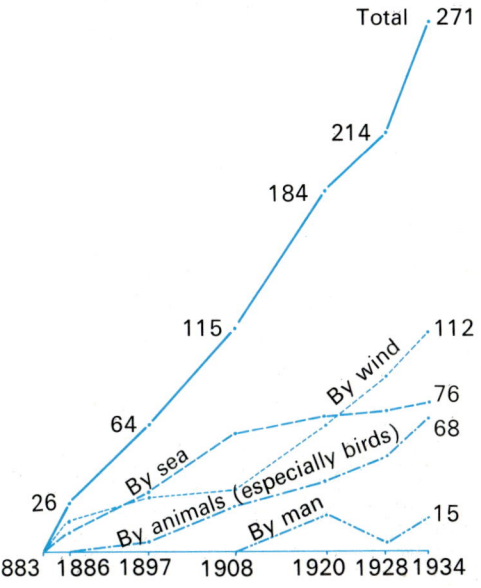

Ou (*Psittirostra psittacea*)  Palila (*Psittirostra bailleui*)

Laysan finch (*Psittirostra cantans*)  No common name (*Psittirostra palmeri*)

No common name (*Psittirostra flaviceps*)  Kona finch (*Psittirostra kona*)

The Hawaiian honeycreepers of the family Drepanididae are astonishingly diversified, not only at generic level but also within a single genus, as is shown here in the beak shapes of different species of the genus *Psittirostra*.

Island species of plants assume a variety of strange and beautiful forms, as with this splendid flower from Tahiti.

Not all insects are carried along passively by the wind. Those capable of flying can cover even greater distances, and as an example of aerial diffusion the case is often quoted of a butterfly which fell onto the deck of a ship at latitude 71°S, in Antarctica.

In similar fashion birds are liable to be carried off course by storms and other forms of air turbulence, ending up far from their intended destination. After the eruption of Krakatoa in 1883, according to observations carried out in that year and continuing until 1919, twenty-seven species of non-migratory birds settled in the surrounding islands.

A fourth equally important method of dispersal, likewise aerial, involves the participation of the flocks of birds which cross the immense expanses of the Pacific in their amazing seasonal migration flights. A classical example is the golden plover which makes an annual journey from Alaska and Siberia to the Hawaiian Islands. In the course of its migration each bird may carry seeds in its digestive tract, attached to its plumage or stuck to the mud on its feet. Insects' eggs and tiny molluscs may also be transported in this manner.

The role played by animals in plant dissemination has long been recognised and many plants are adapted in such a way that they become firmly attached to hair or feathers. Besides, when birds from wet regions nest or simply fly to and fro to drink, mud usually collects around their feet or in their plumage. To verify this theory a group of scientists collected mud samples from various parts of the bodies of birds captured on Christmas Island, south of Java. Embedded in the mud were twenty-one different seed species.

The albatross constitutes a special case for the feet of this bird are covered with an impermeable secretion, so that the seeds enclosed therein are fully protected against sea water and will be dropped intact after a considerable time and often a good distance from their country of origin. Albatrosses cover a large part of the globe in their wanderings.

Not everyone accepts the theory of seeds being carried inside the digestive tracts of animals, and indeed the subject is highly controversial. It has of course been established that certain plants are disseminated through the intermediary of herbivores and frugivores; many such seeds germinate directly in the course of their passage through the intestines. This procedure has an additional advantage by reason of the fact that the seeds are deposited in a well fertilised spot since they are eliminated from the animal's body with the excreta. To determine whether such a germinating process was valid for migrating bird species covering immense distances, experiments were carried out to estimate the length of time required for a food particle to travel through the intestines. It proved to be comparatively short, at most seven and a half hours. Nevertheless, more tests of this nature will be necessary before science can draw definite conclusions for there are a number of other facts which would seem to support the feasibility of long-distance internal transportation. The speed of passage through the digestive tract depends, for example, on how much food is present; apparently the rate is faster if the intestines are full, and vice-versa. Furthermore,

nobody knows what effect the impact of strong winds may have on a bird's digestive processes. It is probable that, as happens under the influence of various forms of stress, the digestive functions are completely paralysed. Finally, it must be obvious that the digestive mechanism works differently in every species.

Another objection to the theory of intestinal transportation is that fruit-eating birds are not known to have strong migratory instincts. Nevertheless, recent studies have shown that they travel much farther than was once believed. It is possible, therefore, that they may indeed be agents of seed dispersal.

What must be remembered, however, is that once they have reached an island, by whatever means, the difficulties of these accidental immigrants, whether plants or animals, are only just beginning. Even after a safe journey the colonisation process may be cut short. Thus the efforts of a plant to establish itself may be prevented for lack of pollinating agents. Then too, although certain factors may favour initial establishment, as in the case of Pacific plants rooted in volcanic soil, the situation of animals is far more complicated, simply because an individual of the same species but of the opposite sex is needed for reproductive purposes. The occasional rare case of an already gravid female perpetuating the species is too remote to be taken into serious consideration.

Quite apart from such factors there must be certain natural preconditions. For a herbivore there must be adequate plant

Many of the islands of the Pacific Ocean are of volcanic origin, the piles of accumulated lava towering many thousands of feet above the ocean floor. Some of them are densely covered with vegetation, the seeds having been carried to these shores by air and water currents, by birds and, where there is evidence of human population, by man.

cover. On a newly emerged island the first plants to appear would be those rooting in rocky soil, followed by small herbaceous species. In favourable climatic conditions a forest might eventually develop. A similar gradual process of consolidation and growth can be envisaged in the case of animals.

Because colonisation of islands is such a haphazard affair, however, the establishment of a healthy plant and animal community must differ from that involved, say, in repopulating a devastated continental region. In the former case entire groups of flora and fauna will be absent. This is why the ecological structure of so many islands is inherently unbalanced.

Some successful examples of colonisation, lending support to such theories, date from recent years, as on Krakatoa, which was destroyed by volcanic eruption in 1883, and on Surtsey, an island which suddenly appeared off the coast of Iceland in 1963. In the latter case various plants and a species of mosquito were found to have established themselves very soon after the island rose from the depths of the sea, and it is now a port of call for several species of birds.

## The beautiful Hawaiian Islands

The lovely islands which make up the Hawaiian archipelago are ideal places for the study of many of the problems relating to island flora and fauna, to which we have briefly referred. Below are some revealing statistics produced by Zimmerman:

| Groups of organisms | Species native to Hawaiian Islands | Number of original immigrant species |
|---|---|---|
| Insects | 3,722 | 233–254 |
| Land molluscs | 1,064 | 22–24 |
| Birds | 70 | 15 |
| Phanerogamous plants | 1,729 | 272 |
| Ferns and other pteridophytes | 168 | 135 |

Although bats are the only mammals that have proved capable of colonising the Hawaiian Islands, marine mammals such as monk seals (seen here basking on the rocky coasts) have managed to find their way to these shores from the American mainland.

Because of their complex structure and isolated situation the Hawaiian Islands are ideal places for the study of evolution and distribution of flora and fauna.

Bearing in mind that the oldest land formations in the archipelago are believed to date back five million years or so, one can work out (dividing five million by the figures in the right-hand column) how frequently colonisation by ancestral forms of new species would have occurred. Thus in the case of insects it is about once every 20,000 years. For land molluscs the interval between each successful settlement is more than 200,000 years; and for birds this figure rises to approximately 300,000 years. This is a surprising result, especially in the disparity between insects and birds, which underlines the importance of wind dispersal.

What happened after the original immigrant species reached the Hawaiian Islands? Study of the pattern of the Hawaiian fauna supplies an answer to this question. As might be gathered from a quick glance at the table on the opposite page, the foreign species that evolved in complete isolation eventually gave rise to new species which are endemic, that is, encountered nowhere else in the world. Thus in the case of insects, 99 per cent of the 3,722 species (grouped in five orders) are considered to be endemic.

One of the most interesting characteristics of island flora and fauna is the so-called adaptive radiation. The first arrivals found themselves in an almost totally uninhabited environment with a wide range of available, easily accessible, ecological niches. This enabled species to evolve in all manner of ways. Such adaptive radiation is particularly striking among insects. According to Jean Dorst, there are in Hawaii four genera which each contain more than one hundred species, and one of these (*Hyposmocoma*) has 216 species. In addition there are ten genera each comprising over fifty species, twenty-four with more than twenty-five species, and forty-seven with over ten species. Consequently, out of a grand total of 3,722 species, 2,963 (about 79 per cent) belong to only eighty-five genera, and the latter make up approximately 22 per cent of the 377 indigenous genera.

This spectacular process has been made possible not only because of the multiplicity of vacant ecological niches but also as a result of the mountainous character of the islands, which serves to separate populations. In the case of terrestrial molluscs, for example, each valley virtually possesses its own species. There are, for example, five genera of Amastridae, comprising 294 different species.

## The extraordinary honeycreepers

The most dramatic example of adaptive radiation, however, is found among the Drepanididae or Hawaiian honeycreepers, a family of birds native to the archipelago. This family is divided into two subfamilies, made up of nine genera and about twenty-two species, a dozen of which are today extinct. This does not take into account other species that may have vanished without leaving any traces behind. The Hawaiian honeycreepers exist in such an astonishing range of forms that the earliest zoologists to study them classified them in three different families. Some were declared to be Fringillidae (finches), some Dicaeidae

Over the years many island species have become extinct and others are still in serious peril. The Hawaiian goose or nene (*Branta sandvicensis*) seemed destined for extinction in the wild but today its numbers are again increasing. It has also bred successfully in captivity and there is a good-sized flock at the British Wildfowl Trust's headquarters at Slimbridge.

Akialoa
(*Hemignathus obscurus*)

Crested honeycreeper
(*Palmeria dolei*)

Akiapolaau
(*Hemignathus wilsoni*)

Pseudonestor
(*Pseudonestor xanthophrys*)

Iiwi
(*Vestiaria coccinea*)

Laysan finch
(*Psittirostra cantans*)

(flowerpeckers) and others Meliphagidae (honeyeaters). Later observations led to their being grouped in a single family probably derived from the Coerebinae (honeycreepers).

It appears that the Hawaiian honeycreepers evolved more or less as follows. The colonising stock must have been similar to the modern subspecies *Loxops virens chloris,* feeding mainly on nectar and supplementing its diet with small insects trapped in the flowers. Later, when the comparatively restricted habitat was completely populated, it apparently deviated from its normal food pattern, giving rise on the one hand to insectivores and on the other to granivores. Then followed a series of further modifications of these forms. Each of the nine resultant genera had their individual food patterns and fed in different types of habitat; and each genus produced a number of species that were basically distinguished from one another by the shape and structure of the bill.

In other respects, apart, that is, from the type of beak and food habits, all Drepanididae are similar in appearance and behaviour, although both subfamilies have different forms of song. Studies in the wild involving the iiwi (*Vestiaria coccinea*), the apapane (*Himatione sanguinea*) and the amakihi (*Loxops virens*) also indicate that these three species, particularly the last, are to some extent territorial by habit. All females lay two or three eggs in spring.

*Facing page:* Despite appearances these six birds are all species of Hawaiian honeycreeper belonging to the same family, Drepanididae. The two belonging to the genus *Hemignathus*, however, are extinct. One of the most handsome of the surviving species is the iiwi (*below*) with bright red plumage and black wings and tail

The extraordinary diversification of the Drepanididae is shown in this family tree, drawn by Dr Dean Amadon. In each of the nine genera the structure of the tongue and the shape of the bill are strictly related to the birds' feeding habits.

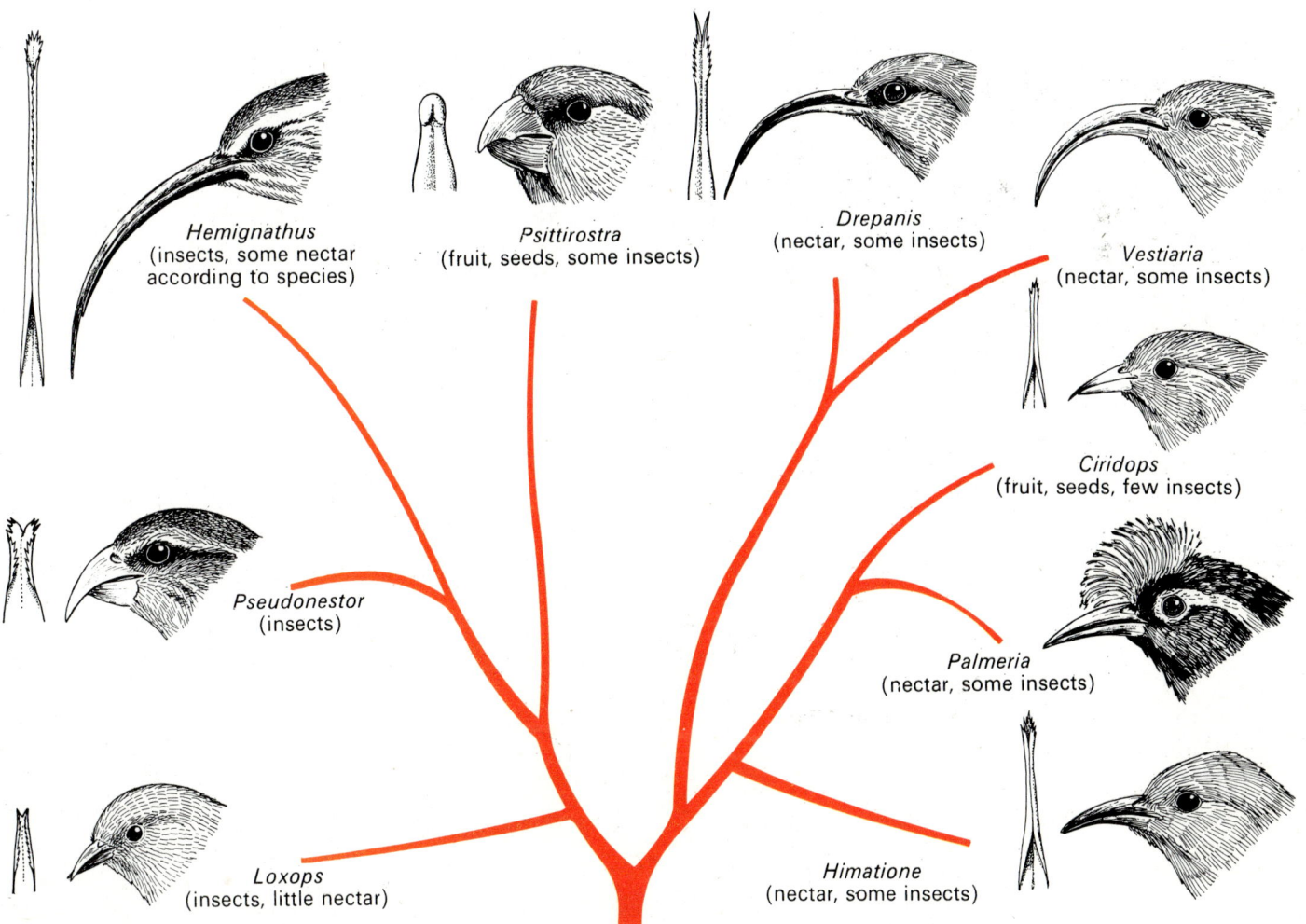

# CHAPTER 108

# The outposts of the Atlantic

The Atlantic Ocean, described by ancient navigators as the Dark Sea, first began to yield up its secrets when hardy Mediterranean seamen sailed past the Pillars of Hercules and set their course southward into the unknown. These were voyages of exploration fraught with danger and excitement, and it is hardly surprising that, used as they were to the tranquil, land-locked waters of the Mediterranean, these earliest ships should cling to the African coasts where they could be sure of finding food, fresh water and shelter from storms. There was no sense in braving the physical and supernatural terrors of the dark, deep waters beyond, arousing the wrath of sea spirits and gods.

One of the most persistent legends concerning the Atlantic told of a vast continent which lay somewhere to the west, with mountain peaks as lofty as those of the Atlas range. This kingdom, possessed of immense natural riches and cradling an advanced civilisation, was known by the name of Atlantis. But one day the ships that regularly traded with this land were compelled to turn back with their cargoes, for Atlantis had vanished overnight, apparently swallowed up by the sea as a result of a catastrophic volcanic eruption or earthquake. Centuries later Phoenician sailors, driven by violent winds from the shores of Africa, discovered islands which the Greeks knew as the Isles of the Blest or Fortunate Islands. Although Atlantis was never located, the Fortunate Islands proved to be real. The elder Pliny wrote of 'Canaria, so called from the multitude of large dogs (*canes*)'. Arab seamen landed there in A.D. 999 and navigators from various European nations reached the islands in the 13th and 14th centuries. Eventually, in the late 15th century, the Spaniards conquered and claimed them for the Crown under the name Islas Canarias.

*Facing page:* Two-thirds of the land surface of the island of Tenerife consists of an immense volcanic dome with undulating plains of pumice and lava and dominated by the huge Pico de Teide. The central cone has a double summit, the higher of which, El Pitón, rises to 12,172 feet.

Experts are still not agreed as to how the Canary Islands originated nor how to explain the existence of so many endemic species of plants and animals. The archipelago would seem to be of volcanic origin, emerging from the depths as a result of an immense underwater eruption in the course of which huge quantities of volcanic material, chiefly acidic and basaltic rock, accumulated. The majority of authors affirm that the islands were never attached to any continent and that their flora and fauna must have reached these shores by passive methods during the Miocene epoch, evolving in the course of the Pliocene and the Quaternary period, thus giving rise to innumerable endemic forms which still exist today.

Nevertheless, the discovery on Tenerife of a huge fossil land tortoise (*Testudo burchardi*), weighing more than 300 lb and dating from the Tertiary period, does seem to support the contrary assertion that the Fortunate Islands were once joined to the African continent, although separated many millions of years ago. It lends plausibility to the theory advanced in the 19th century by Francisco Quiroga and later supported by S. Calderón, Hernández Pacheco and Vidal Box. These Spanish geologists claim that the islands broke away as a consequence of the immense tectonic upheavals which produced the Alps, the Pyrenees and the Atlas mountain chains and which thus shaped the Atlantic island archipelagos.

These tectonic disturbances which continued through the Miocene and Pleistocene, unleashed wide-scale volcanic activity with the result that the primitive crystalline foundation was overlaid by eruptive material, suggesting a volcanic origin for the islands. A. Wegener may be closest to the truth with his theory of continental drift which postulates the separation of the islands from the African land mass in the form of fragments of sial (a combination of silica and alumina) carried on a flow of sima (silica and magnesia).

The fertile deposits of lava, combined with a warm climate which, although tropical, is modified by ocean currents, have stimulated the growth of a characteristically luxuriant plant cover which has led botanists to group the Canaries with the Azores, Madeira, Cape Verde Islands and Ilhas Selvagens (all Portuguese). To the north the Azores form a bridge with the Holarctic region while to the south the Cape Verde Islands are more closely linked with the Ethiopian region.

## A multitude of endemic plants

In all the islands of the Atlantic Ocean the climate is temperate, influenced by the trade winds and characterised by rare, irregular rainfall. The vegetation is varied and in most cases differs from island to island. Broadly speaking, the landscape encountered in the Canaries is found only on Madagascar, in Australia and on some Polynesian islands. It comprises more than 500 species of endemic plants, compared, for example, with less than 100 species on Madeira.

The most typical plant community is the laurel forest which, with the cactus-like euphorbias and legendary dragon trees

(*Dracaena draco*), together with ferns and heather, give the landscape an unusual aspect. In addition to these, familiar species include the Canary pine (*Pinus canariensis*) as well as dense thickets of gorse and juniper in which a large number of birds and reptiles find refuge. Differences in altitude, significant over such a comparatively small area, are partially responsible (perhaps here more than anywhere else in the world) for a truly remarkable diversity of plant life.

At sea-level plants adapted to very dry conditions, such as euphorbias, heather, broom and juniper, show some similarity to the vegetational cover of the neighbouring Sahara coast. At 1,000 feet the euphorbias tend to thin out and give way to pines and related forms. Here may be found a number of very ancient dragon trees to which local people attribute miraculous properties. The laurel forest, perpetually enveloped in cloud, extends between the 1,650- and 5,000-foot marks. The warm, humid trade winds blow from the north-east and as they strike against the peaks give rise to persistent mists which shroud the upper parts of the forest and keep it extremely wet. Tests have shown that more than 120 inches of water may collect under the trees, compared with a maximum of 40 inches on open ground. Above the laurels are fields of gorse and juniper and, as the soil becomes progressively drier with altitude, pines and undergrowth comprised of broom, thyme and cistus. These shrubs are to be found in places up to a height of about 8,000 feet. Heather and broom continue up to an altitude of 10,000 feet or thereabouts, after which the flora is characteristically alpine in general appearance.

The Canary Islands therefore exhibit an extraordinary diversity of habitats, from rocks and huge piles of debris on the shores to regions of almost perpetual snow in the high mountain zones. Between these extremes are areas of semi-desert, flatlands covered with bushes and shrubs, dry pinewoods and tropical rain forest—and all this within a total space of some 2,900 square miles. It is hardly surprising, as a result, that the naturalist should find, alongside characteristic steppe and African desert species, plants that are typical of temperate Europe as well as a large number of endemic forms which are completely unknown elsewhere.

## Animals of the Canary Islands

Because of their proximity to the African mainland the Canaries are, of all the Atlantic islands, the richest in wildlife. Just the same, there are no true freshwater fishes, snakes or indigenous terrestrial mammals here; and batrachians are represented only by the St Anthony's frog. The sole indigenous mammals are bats and (on the coasts) the occasional monk seal. All the other mammals found in the archipelago—mice, rats, hedgehogs, rabbits, goats, camels, dogs etc—have been accidentally or deliberately introduced by man fairly recently.

The ornithologist travelling to the islands by sea will be fascinated by the masses of birds, including gulls, puffins, petrels and albatrosses, which crowd the cliff ledges. Although

Former geographical distribution of Meade Waldo's oystercatcher.

**MEADE WALDO'S OYSTERCATCHER**
(*Haematopus ostralegus meadewaldoi*)

Class: Aves
Order: Charadriiformes
Family: Haematopodidae
Total length: 16½ inches (42 cm)
Wing-length: 10-10¼ inches (25-26 cm)
Length of bill: 3-3¼ inches (7.7-8.1 cm)
Diet: molluscs and crustaceans
Number of eggs: probably one
Incubation: not known

Plumage dull black, except for lower inside parts of primary wing feathers which are white. Iris bright red. Bill of female usually somewhat longer than that of male. Species believed to be extinct.

such sea birds alight on the islands proper for nesting purposes alone, the period which they are compelled to spend on dry land is in fact the most critical of their life, for it is then that they are exposed to extreme danger as a result of man's insatiable greed and cruelty.

Plump puffin chicks, before they are capable of flying, are regarded locally as great food delicacies. In September the puffin colonies are ransacked by hunters, accompanied by ferrets and dogs, who denude nests of eggs and fledglings. The flesh of thousand upon thousand of small birds is salted and preserved in wooden barrels. It goes without saying that such depredations, to which both Spanish and Portuguese authorities have always turned a blind eye, pose a serious threat to the future of the species. Conservationists have protested at this indiscriminate destruction, but to little avail.

Unfortunately no measures of protection can now save Meade Waldo's oystercatcher (*Haematopus ostralegus meadewaldoi*), which has already disappeared, leaving behind a minimum of information about its life habits. Similar to the common oystercatcher of European shores, it differed in outward appearance by reason of its uniformly black plumage which enabled it to remain perfectly camouflaged among the lava rocks of its environment. The last representatives of the subspecies were killed on the coasts of the Canary Islands and today there are only stuffed specimens in some of the world's museums.

All we know of this oystercatcher is that it wandered along the beaches, flitting from bay to bay and island to island in search of food. On shores battered by high waves, stepping gracefully over the rocks of basalt, it would hunt the tiny crustaceans and molluscs which constituted its diet. Breeding habits remain a mystery for neither scientists nor local fishermen record having seen a nest. The female is believed to have laid a single egg in June, for some of the older islanders recollect having seen groups of three of these birds towards the end of August.

Some naturalists have challenged the very existence of Meade Waldo's oystercatcher, alleging that the various sightings must have concerned abnormal forms of the African black oystercatcher (*Haematopus moquini*) which somehow strayed to the Canaries. Other ornithologists suggest a melanistic form of the European oystercatcher. Such claims lack foundation for there is no evidence of these birds having been aberrant individuals.

Although everything points to the fact of this subspecies being extinct, there is just a remote chance that one day a surviving pair may be found on an unexplored rock or isolated cliff ledge. It would not be the first time that an animal given up for lost unexpectedly reappeared.

## Birds of plain, wood and mountain

Striking inland from the coast in search of other typical birds of the islands, sooner or later one is bound to see, on strips of dry land where vegetational cover is sparse, enormous flocks of trumpeter bullfinches. These birds give out an incessant and

monotonous call note, keeping their distance and often vanishing from view, thanks to their astonishingly effective camouflage. This species, which has recently been sighted in Spain, near Almeria, builds its nest in rock crevices.

Rocky terrain is also the favoured habitat of houbara bustards, grouse, cream-coloured coursers and a number of other species originating in desert regions. The houbara bustards and sand grouse are, however, endemic subspecies nearing extinction.

Particularly striking are small birds which peck for food among the stones and pebbles. These are Berthelot's pipits (*Anthus berthelotii*), an indigenous species inhabiting the most arid parts of the archipelago. Food consists of the seeds of herbaceous plants and, above all, insects which the birds hunt tirelessly on foot. The pipits seldom perch on trees for they are much more at home on the ground, preferring to seek safety by running away and only taking wing as a last resort. The nest is built beside a large stone or in the shelter of grass and shrubs. Two, three or four eggs are laid in March, the female being an expert in deception. Should her brood be endangered she will expose herself to the enemy and try to divert the latter's attention with piercing chirps and twists of the body, designed to give the impression that she is badly injured.

The steepness of the slopes and the crumbling texture of the basaltic rocks, combined with the torrential rains of the sub-tropical regions of the islands, have resulted in the formation

Meade Waldo's oystercatcher, now feared extinct, was a beautiful bird whose plumage blended perfectly with the basaltic rocks of its environment.

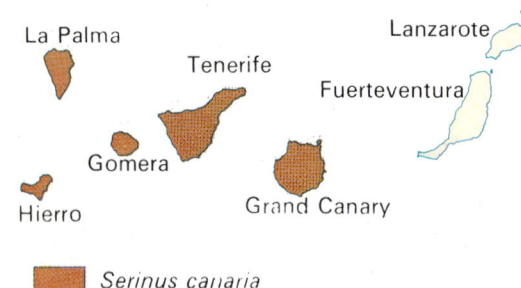

Geographical distribution of the canary (*Serinus canaria*) and the Canary Islands chaffinch (*Fringilla teydea*).

---

**CANARY**
(*Serinus canaria*)

Class: Aves
Order: Passeriformes
Family: Fringillidae
Total length: 5 inches (12.5 cm)
Wing-length: 2¾ inches (7.1 cm)
Diet: seeds and fruit; also insects, especially during rearing of chicks
Number of eggs: 4-5
Incubation: 15 days

Plumage generally greenish, back grey-green, lower part of belly white. Flanks speckled, becoming bright yellow on chest. Female is similar in colour to male but less brilliant.

**CANARY ISLANDS CHAFFINCH**
(*Fringilla teydea*)

Class: Aves
Order: Passeriformes
Family: Fringillidae
Total length: 6¼ inches (16 cm)
Wing-length: 3½-4¼ inches (8.9-10.7 cm)
Diet: insects and seeds, mainly those of Canary pine
Number of eggs: 2
Incubation: 14-15 days

Upper parts beautiful azure blue, white ring around eye. Wings black, the feathers fringed with white. Underparts almost white, but breast bright blue. Colour of female much drabber, a mixture of blue-green and grey, with less prominent ring around eye.

---

of immense ravines, usually bordered by luxuriant plant growth. Here, as in the coastal areas, there are many interesting birds, such as the extremely rare laurel pigeon (*Columba junoniae*), which, as its name suggests, is also an inhabitant of the laurel woods where it feeds on berries. The laurel pigeon's nest, frequently situated in a deep, inaccessible gorge, is constructed of small roots and slender branches. The female lays only one egg, which is surprising in view of the fact that pigeons customarily lay two. This egg is generally laid at the beginning of May, although the timing depends to a large extent on the weather conditions.

The laurel pigeon spends much time searching for seeds and fruit underneath trees, and has developed very strong legs for this purpose. It moves easily and rapidly over the ground, rather like a partridge.

Another representative of the Columbiformes which is found both in the Canaries and on Madeira is the Trocaz pigeon (*Columba trocaz*), nesting in clumps of heather and among giant ferns, 10-20 feet above the ground, as well as on laurels. The nest of this species is similar to that of the European wood pigeon although it gives the impression of being much more solidly constructed. The two parents take turns to incubate the single egg, which may be laid at more or less any time of year, except, apparently between September and November.

Other Columbiformes found in the archipelago are the Canary pigeons (*Columba livia atlantis*) which nest among the cracks and hollows of cliffs, and turtle doves which keep up their gentle cooing in park and woodland.

## Cagebird in the wild: the canary

Another bird which shares the habitat of these pigeons but is a member of the finch family (Fringillidae) is the canary (*Serinus canaria*), close relative of the serin (*Serinus serinus*). It is well known in its domesticated guise as one of the most popular of cagebirds but few visitors to the Canary Islands readily recognise it in the wild.

It was towards the end of the 15th century that an Andalusian bird-keeper brought back to Spain the first wild canaries to be seen on the European mainland. They and their progeny proved to be easily tamed and were soon being sold as delightful pets. In due course their reputation had spread to other parts of Europe and they rapidly became household favourites everywhere, in a variety of shapes and colours. Such canaries, however, bear only the faintest resemblance to their wild relatives, for the latter have drab green plumage.

These wild birds, found in the Canary Islands, the Azores and Madeira, behave much like other finches. Except during the breeding season, they form large flocks which fly to and fro in quest of food. This consists in the main of seeds and other vegetable matter. Sexual activity occurs early in the year. From January onwards the males devote themselves to song and indulge in nuptial flights to win the favours of the opposite sex.

The precise time for egg-laying varies from one island to

another. Thus on Tenerife it occurs much earlier than on Grand Canary. Timing also depends on altitude. At sea level a female will lay her eggs in March but in the high mountain regions not until June or July. Once the territory of the male is staked out, preparations for nest-building can commence. The site will usually be some 12-20 feet above ground on the branches of plantains, laurels and pines, but also among flowering heathers, cistus and euphorbias. Materials include vegetable fibres, grass and cistus leaves, and it is often lined with lichen, hair and down. The female lays four or five eggs, blue-green with flecks of violet, although sometimes the shell is completely white. The male plays no part in the fifteen-day incubation, displaying no interest in the brood until hatched, but then assiduously cramming his offspring with insects.

## Chaffinches and chats

The Canary Islands chaffinch (*Fringilla teydea*) is another fascinating bird which inhabits the pine forests of Tenerife and Grand Canary, each island having its own distinctive subspecies. It is a bold, trusting bird, threatened only by the comparatively rare sparrowhawk, and will allow humans to come quite close before flying off. Unfortunately, this confidence has proved a disservice which may result in it ending up as a stuffed specimen in some foreign museum.

In addition to pine forests, the Canary Islands chaffinch frequents laurel woods, clumps of heather, undergrowth and, occasionally, desolate moorland. It eats large quantities of insects, including beetles, butterflies and other invertebrates, both on the ground and in the air. But the basic constituents of its diet are seeds, especially those of the Canary pine.

Canary Islands chaffinches form pairs in April and May, both

The canary, today best known as a cagebird, is found in its wild state only in the Canary Islands.

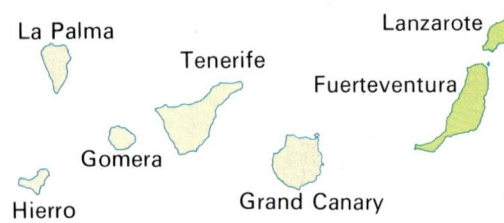

Geographical distribution of the Canary Islands chat.

---

**CANARY ISLANDS CHAT**
(*Saxicola dacotiae*)

Class: Aves
Order: Passeriformes
Family: Muscicapidae
Total length: 5 inches (12.5 cm)
Wing-length: 2½ inches (6.1-6.3 cm)
Diet: insects
Number of eggs: 4-5
Incubation: about 13 days

Upper parts handsome grey with a white mark on secondary remiges. Underparts white except for bright orange-red patch on breast. Female more drab, no colours on breast.

---

*Facing page:* A representative selection of birds endemic to one or more of the islands of the Atlantic Ocean.

partners collaborating in building the nest, which is situated on a lateral branch close to the trunk, anywhere from 3-30 feet from the ground. When complete it is a small masterpiece of interlaced pine needles, moss and lichen, often lined with hair and down. The female lays two eggs, from four to six days apart, around mid-June. The male feeds his mate during the two-week incubation period. Both adult birds then provide the babies with insects and pine seeds already moistened in the crop.

The Canary Islands chaffinch breeds only once a year and the offspring, for some unknown reason, eventually build their nests and breed in their turn much later in the year than the parents, with the result that broods may sometimes be found up to the beginning of September.

Much more common than the Canary Islands chaffinch, with four different subspecies inhabiting the islands of Tenerife, La Palma, Hierro (all in the Canaries), the Azores and Madeira, is the chaffinch (*Fringilla coelebs*).

Another very interesting endemic species is the Canary Islands chat (*Saxicola dacotiae*), a delightful bird which is found in the eastern islands of the archipelago and behaves much like related continental chats. The nest will generally be placed in a hole on the ground or in a rock cleft. Here the female normally lays five eggs during January. The fledglings grow rapidly and can already fly by February.

Among other European birds commonly encountered in the Canary Islands are the woodcock (an inhabitant of the laurel woods) an unusual subspecies of the great spotted woodpecker, the pallid swift, the tawny owl and the long-eared owl, as well as various partridges and quails. Then too there are the Canary tits, with subspecies on most of the islands, although on La Palma nobody has yet found a nest containing eggs. The list of birds also includes tiny wrens, lesser short-toed larks, blackcaps, spectacled warblers, robins, wagtails, blackbirds, shrikes, linnets, sparrows, crows and choughs.

There are a few diurnal birds of prey in the Atlantic islands. Hunting close to the shores are a subspecies of peregrine falcon, the Barbary falcon (*Falco peregrinus pelegrinoides*) and, occasionally, an Eleanora's falcon. Fish eagles sometimes nest on the high cliffs, and red kites, although only on rare occasions, explore the mountainsides. Steep rock faces are also nesting sites for Egyptian vultures and a local species of the common buzzard (*Buteo buteo insularum*). The list is completed by kestrels and sparrowhawks. The latter are comparatively rare but the various subspecies are far more diversified than the continental forms of these raptors.

## Reptiles and amphibians

During the Tertiary period North-west Africa, today an immense tract of desert, was covered with dense vegetation. This was the home of numerous species of reptiles which in form represented an intermediate stage between the giant reptiles of the Mesozoic and those with which we are familiar today. Incapable of adapting to surroundings that were rapidly being

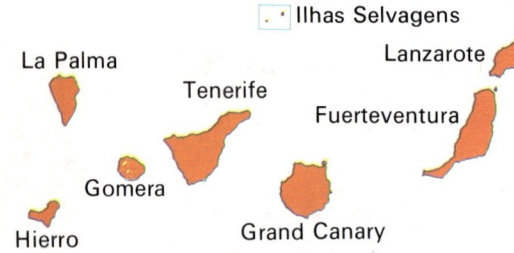

Geographical distribution of Berthelot's pipit.

transformed, many of these species vanished; others were perhaps carried off on fragments of the continent which may have become detached at the time of the Alpine folding. The species thus isolated produced a large number of local races but many of these duly became extinct. This is what happened to the ground tortoises and to three varieties of lizards whose fossil remains have since been discovered.

When the Spanish conquistadors landed in the Canaries they must have found large numbers of lizards but, as a result of indiscriminate hunting, most of them probably disappeared in the course of the 16th century. A few, nevertheless, did manage to survive, notably Simony's lizard (*Lacerta simonyi*). Unfortunately this species too was virtually wiped out in the cause of science by over-zealous zoologists. The last representatives of the species clung to a few small islets (Roques del Salmor) where they lived in symbiosis with a colony of sea birds. The twenty or thirty lizards occupying this restricted site measured about 3 feet long and apparently lived on the excrement, eggs and chicks of their bird hosts. For centuries the little community remained intact simply because it was impossible to reach them; but in the end they were decimated at the behest of trophy hunters who offered vast sums of money for their capture. Today a few survivors of the subspecies *Lacerta simonyi stehlinii* live on Grand Canary. They are small and dark, almost black, with a white belly.

Another more familiar species, best represented on the western islands, is Gallot's lizard (*Lacerta galloti*), which roams many habitats with the exception of the rain forests and laurel woods, where perpetual belts of cloud and mist prevent the sun breaking through. Gallot's lizards are likely to be encountered both at sea level and at considerable altitude. On Tenerife, for example, they have been sighted at 10,500 feet. Their food consists principally of vegetables and fruits, and they often create havoc in tomato plantations and vineyards. For this reason they are unmercifully hunted by farmers and market gardeners who often make use of poisoned bait. But their diet is broadly based and also includes large numbers of insects, small vertebrates and, on occasion, carrion.

During the breeding season the backs of the males turn deep black while beautiful blue marks appear on the flanks. The Spanish herpetologist Alfredo Salvador, who has made a detailed study of their habits, believes them to be strongly territorial for there are frequent combats between rival males, although dissuasive rituals prevent them ending fatally. Fights are sometimes avoided altogether after a series of intimidating manoeuvres in which the two combatants take up a stance on a piece of open ground, swell their throats and let out a sequence of grating cries of warning. The females, on the other hand, with their drab breeding colours, remain in hiding. This species would seem to have originated on Tenerife, later colonising Gomera, Hierro and La Palma, where the subspecies are all somewhat smaller.

The Canary's Island green skink (*Chalcides viridianus*) has a wide range of distribution throughout the archipelago but does

---

**BERTHELOT'S PIPIT**
(*Anthus berthelotii*)

Class: Aves
Order: Charadriiformes
Family: Motacillidae
Total length: 5½ inches (14 cm)
Diet: seeds and, above all, insects
Number of eggs: 2-4
Incubation: about 12 days

Back ochre-grey, belly white and prominently speckled. Easily distinguished in flight by two white outer rectrices. Common on almost all islands and islets in Canaries, especially on open land, volcanic regions and semi-desert zones.

not clash with other lizards because of its preponderantly nocturnal habits. Its colour is usually black with copper overtones on the back. By day it remains concealed beneath stones or in bushes, although sometimes it may be seen basking peacefully in the sun. At the slightest noise, however, it makes off; but because its legs are atrophied and thus of little use, it moves in an undulating fashion, rather in the manner of a snake. The skink is ovoviviparous, the female giving birth to two babies in the spring. The latter are as agile at birth as their parents, although they only reach adult stature at the age of three or four years. The species is territorial and social relationships are very elaborate. Individuals of lower rank wave their tails vigorously when meeting a dominant, so avoiding combat.

In the Cape Verde Islands there is an extremely interesting analogous species, the Cape Verde skink (*Macroscincus coctaei*), which has posed a number of zoogeographical problems that are hard to resolve. The reptile, in fact, is only found on the islets of Branco and Raso. What is more, the skink is in every way similar to a giant skink which was until quite recently an inhabitant of some small islands in the Indian Ocean. Apart from this surprising feature of distribution, other remarkable characteristics of the Cape Verde skink are its diet, which is exclusively vegetarian, and its prehensile tail–highly strange for an animal which lives in a habitat where there are no shrubs or trees. The conclusion must be that long ago this species lived in completely different surroundings where it evidently was arboreal in habit.

Although the marsh frog (*Rana ridibunda*) is common enough, it is an introduced species. The only indigenous amphibian of the Canaries appears to be the Mediterranean tree frog (*Hyla meridionalis*). This is a fairly recent arrival which has adapted in remarkable fashion to island life. Given its dependence on water, it is the only poikilothermic vertebrate living in the laurel woods, where its population reaches explosive proportions; and it is abundant too in other biomes, including cultivated land. The English ecologist Hugh B. Cott, who carried out a count of the species in banana plantations, produced the astonishing figure of approximately one million frogs to the square mile. According to Cott there was at least one animal in every tree and there was no need to move about for food. Despite their numbers the frogs are a boon to growers for they devour immense quantities of parasites.

Gallot's lizard is found in many parts of the Canary Islands, apart from the laurel forests which are perpetually shrouded in mist. Although it often appears in plantations and vineyards it is useful in the sense that it destroys enormous quantities of insects.

## A vanishing world

It has been pretty well established that the first people to inhabit the Canary Islands were Cro-Magnon hunters driven southwards from Central Europe during the Upper Palaeolithic period by advancing ice sheets. Towards the end of the Neolithic the archipelago was settled by the Mechta el Arbi or Berber people. Their fusion with the earlier inhabitants of the islands gave rise to the Guanches, a name strictly applicable to the primitive inhabitants of the western islands but later used to describe all the pre-Hispanic occupants of the archipelago.

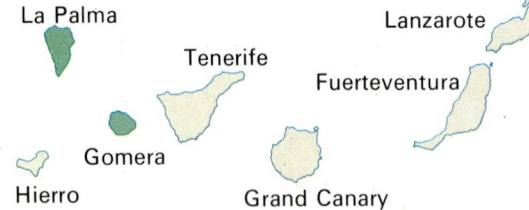

Geographical distribution of the laurel pigeon.

The Guanches were farmers and breeders of livestock and their way of life undoubtedly had an effect on the wild landscape of the islands; but although they apparently built a complex network of canals, the primitive character of their plantations and the relatively small numbers of their livestock cannot have had a seriously disturbing influence on the local balance of nature. It was the Spaniards, conquering the islands at the end of the 15th century, who began to despoil the land. Their deforestation and overgrazing activities brought about a complete, tragic transformation of the original ecosystem.

The repercussions on local flora and fauna were to continue into the 20th century. The recent disappearance of Meade Waldo's oystercatcher and the large Simony's lizard are two classic examples of the way man has tampered with the delicate equilibrium of island wildlife. Only a few years ago, too, Egyptian vultures still played a significant part in pest control, for in addition to feeding on carrion they hunted reptiles and insects. But indiscriminate spraying of DDT to combat locust plagues decimated the birds to such a degree that one can travel many miles nowadays without sighting a single pair.

Reports by naturalists who have lately visited the islands are alarming. Cuyas Robinson writes that the laurel woods have completely vanished on some islands, including Grand Canary; and it is clear that land clearance schemes, the haphazard use of pesticides, the destruction of undergrowth and the human population explosion in the islands will sooner or later bring about the complete destruction of the natural scene. Even in officially protected zones such as the Teide National Park on Tenerife, pines have replaced the original flora.

It is vital that an effective conservation policy should be worked out so that what remains of the original biomes can be saved, whether they be deserts, steppes, forests or mountains: and only in this way can we be sure of rescuing from extinction such rare species as the laurel pigeon.

## Annual meeting point for sea birds

The small volcanic islands and islets of the Atlantic, some of them more than 1,250 miles from the nearest continent, are meeting places for enormous breeding colonies of sea birds. But the majority of them do not accommodate any mammals or indigenous ground birds. Such is the case on the island of Ascension, halfway between Africa and America, and St Helena, situated more than 1,800 miles from the shores of the New World and just over 1,100 miles from Europe.

The lonely islands of Tristan da Cunha and Gough, which lie farther south than the Cape of Good Hope, are also important breeding sites for many sea birds but, unlike those mentioned above, the islands also shelter a number of indigenous land birds such as the flightless Inaccessible Island rail (*Atlantisa rogersi*), two species of finch of the genus *Nesospiza* and a starling. But by far the most interesting bird of these islands is the great shearwater (*Puffinus gravis*), a typical sea bird which visits dry land only for nesting.

---

**LAUREL PIGEON**
(*Columba junoniae*)

Class: Aves
Order: Columbiformes
Family: Columbidae
Total length: 14-16 inches (35-40 cm)
Wing-length: 8-8¾ inches (20.5-22 cm)
Diet: principally seeds and fruit
Number of eggs: one
Incubation: about 19 days

Head and upper parts blue-grey; sides of neck and nape show metallic green reflections turning to purple at base of neck and on scapulars. Wings greenish-brown and underparts pink. Tail pale grey, almost white at tip.